BRIDGING
English Language Learners to GED® Test Prep

TEACHER'S GUIDE

Mathematical Reasoning

Lia Conklin Olson

Bridging English Language Learners for GED® Test Prep: Mathematical Reasoning
ISBN 978-1-56420-076-1

New Readers Press
ProLiteracy's Publishing Division
308 Maltbie Street, Suite 100, Syracuse, New York 13204
www.newreaderspress.com

Printed in the United States of America
10 9 8 7 6 5 4

Proceeds from the sale of New Readers Press materials support professional development, training, and technical assistance programs of ProLiteracy that benefit local literacy programs in the U.S. and around the globe.

Editor: Laura McLoughlin
Editorial Director: Terrie Lipke
Designer: Cathi Miller
Technology Specialist: Maryellen Casey

CONTENTS

CONTENTS

MATH SENSE 3: FOCUS ON ANALYSIS

Note: This book refers to lessons and page numbers in the 2015 editions of *Math Sense*.
If using earlier editions, page numbers and lessons may not match.

About *Bridging English Language Learners to GED® Test Prep*

Why *Bridging*?

If our GED® classrooms were filled with students who came to us with solid foundations in math skills and conceptual understanding alongside a high level of English proficiency, there would be no need for this book. Providing students with guidance as they complete the practices in the New Readers Press GED preparation materials would be enough. The reality is, of course, that by and large our students come to us for GED preparation with math skills far below that. When we add to that the increasing number of English language learners (ELLs) with language development needs, preparing our learners for the GED test becomes quite a challenge. The good news is we have a job to do—and an important one! The better news is that this teacher's guide can help us bridge our ELLs from where they are to where they need to be. A wise man (Lev Vygotsky) once hypothesized that there is a zone of proximal development in which a knowledgeable other (that's you) could push students from where they are to greater levels of knowledge. It is with this in mind that *Bridging* was born.

ELLs come to us with an abundance of skills, and it is up to us to attach trusses of learning to those skills to provide access to GED preparation even when their math skills are "too" low and their language skills are "too" underdeveloped. As the knowledgeable others, we can provide learning opportunities that bridge these gaps to open opportunities for students to learn GED content. This teacher's guide is intended for instructors of students with basic math skills who need to develop deeper conceptual understanding of math and build their English language skills. That said, we all know that students with low math skills and English language skills come through our doors eager to do what they need to do to get their GEDs. Although it is unlikely that these students will pass the GED test within a short span of time, it is possible that providing these trusses of learning will not only allow them access to GED test prep materials but may also increase their mathematical competency and language development sufficiently so that by the time they take the test, they may have the skills they need to be successful. So open up this teacher's guide and embrace your ELLs and their mathematical and language needs. With a little prep, you can begin to bridge their journey to the GED test.

What's in *Bridging*?

This guide provides:

- Lessons that correspond directly to and support each unit of the New Readers Press *Math Sense®* series. Each *Bridging* lesson covers one unit.
- Skills-based questions that focus each lesson
- Learning goals that are clearly defined and aligned to GED assessment targets
- Instructional strategies that support students' knowledge building, problem solving, vocabulary development, and math application skills
- References to pre-GED resources and specialized lessons in the *Math Sense* books that bridge learners' prior knowledge to GED content
- Example instructional activities that model instructional strategies
- Vocabulary development strategies and word lists

Materials Needed to Use this Book

This teacher's guide follows the units within each book of the *Math Sense* series. This book also links the material covered in each *Math Sense* unit to other corresponding New Readers Press materials that will build on the content required on the GED test. These books are available to purchase at newreaderspress.com.

This guide is based on these New Readers Press materials:

- *Math Sense® 1: Focus on Operations*
- *Math Sense® 2: Focus on Problem Solving*
- *Math Sense® 3: Focus on Analysis*

Corresponding materials:

- *Pre-HSE Core Skills in Mathematics*
- *Scoreboost® for the GED® Test: Fractions, Decimals, Percents, and Proportions*
- *Scoreboost® for the GED® Test: Algebraic Reasoning*
- *Scoreboost® for the GED® Test: Measurement and Geometry*
- *Scoreboost® for the GED® Test: Graphs, Data Analysis, and Probability*
- *Pre-High School Equivalency Workbook: Math 1 (Whole Numbers, Decimals, Fractions, Percents, and Measurement)*
- *Pre-High School Equivalency Workbook: Math 2 (Algebraic Thinking, Data Analysis, and Probability)*

How Does *Bridging* Work?

This graphic illustrates the relationship between each unit's Learning Goals and the Sample Instructional Support Strategies.

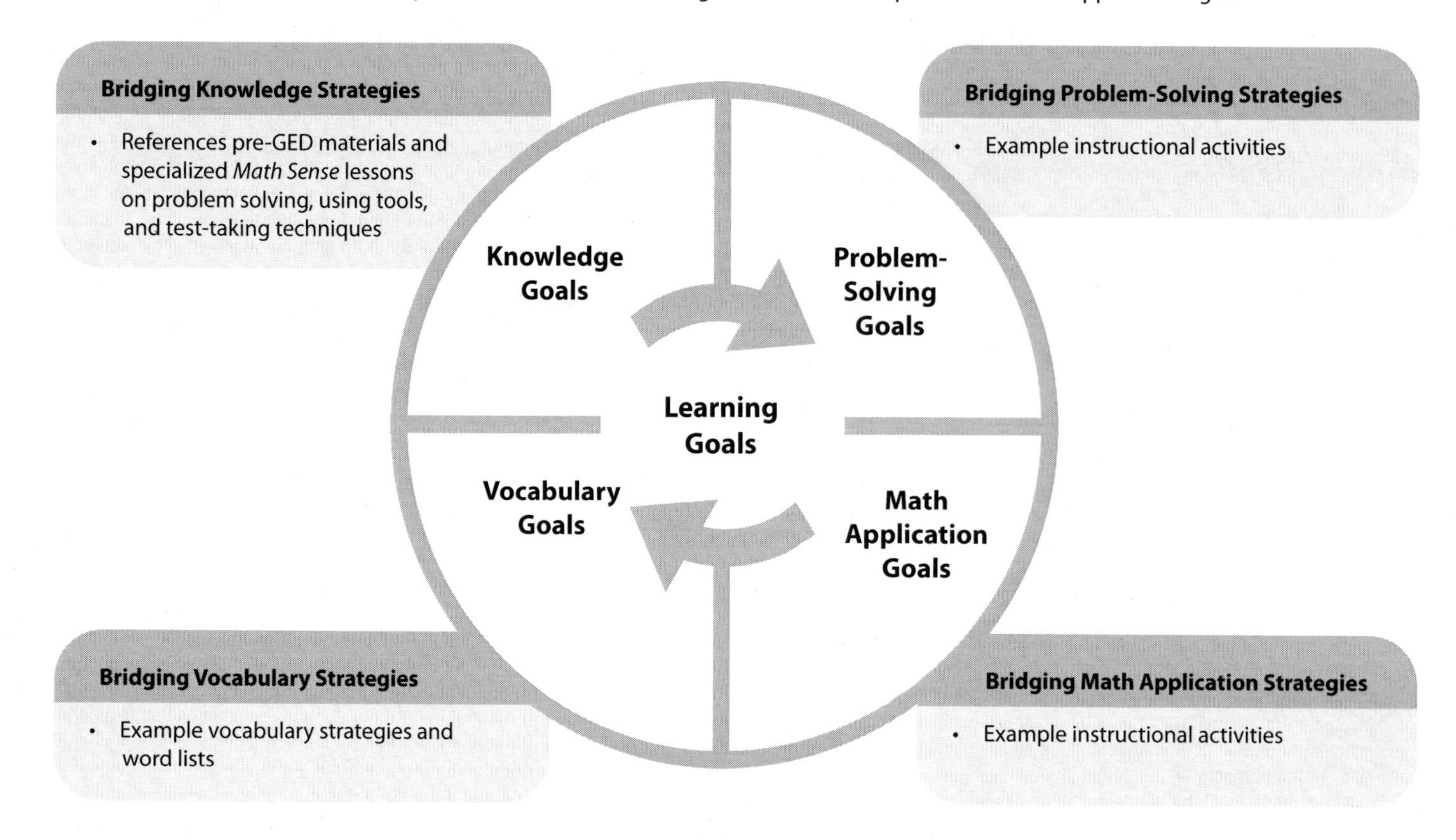

➢ Skills-Based Questions

Each *Bridging* lesson begins with Skills-Based Questions that focus the lesson on specific skills within the mathematics content of the corresponding *Math Sense* unit. Providing an overview of the skill development students can expect in the form of a question allows students to assess themselves once the lesson is completed. Here are two examples from the *Bridging* lesson that supports *Math Sense 2*, Unit 2, Section 1: Algebra Basics:

- What are three mathematical statements used in algebra? How do you translate their symbols into words?
- What does it mean to evaluate an expression? How do you evaluate expressions?

➢ Learning Goals

Learning Goals, the knowledge and skills students will be able to demonstrate upon completion of the lesson, are listed in student-friendly language for four categories of learning: Knowledge, Problem Solving, Vocabulary, and Math Application. All learning goals are directly tied to the GED Assessment Targets for Mathematics. (You can find these targets online in the Assessment Guide for Educators at gedtestingservice.com.) Furthermore, these learning goals are directly aligned to specific parts of each unit in the *Math Sense* books. For example, each unit in *Math Sense* books 1–3 contains four to 10 lessons as well as several specialized Problem Solver, Tools, and Test Taker lessons. *Math Sense* units with seven or fewer parts are supported by one *Bridging* lesson. *Math Sense* units with more than seven parts are supported by two *Bridging* lessons (Section 1 and Section 2). *Bridging* refers to each part of the *Math Sense* units as Part 1, Part 2, etc. following the order they appear in the text.

Here is an abbreviated sample of the Learning Goals from the *Bridging* lesson that supports Unit 2: The Basics of Algebra, Section 1:

	Learning Goals	GED Assessment Targets GED
Knowledge Goals:	1. Describe, in words and symbols, different ways to show the relationship between numbers and variables. *(Part 1)* 2. Describe how to evaluate an expression. *(Part 2)* 3. Explain how to simplify expressions. *(Part 3)*	A.1.e A.1.g A.1.j A. 1.i
Problem-Solving Goals:	1. Translate expressions, equations, and inequalities from symbols to words and from words to symbols. *(Part 1)* 2. Evaluate expressions by replacing variables with given values. *(Part 2)* 3. Simplify expressions by grouping like terms and numbers. *(Part 3)*	A.1.e A.1.g A.1.j A. 1.i
Vocabulary Goals:	1. Define key mathematical terms. 2. Determine the meaning of unknown vocabulary using context clues, word forms, and parts of speech. 3. Apply new vocabulary to mathematical tasks and discussions.	
Math Application Goals:	1. Apply knowledge of the basics of algebra to solve real-life math problems. *(Parts 1–3)* 2. Defend math applications and reasoning to others. (*Parts 1–3)*	A.1.e A.1.g A.1.j A. 1.i Q.1.c

Sample Instructional Support Strategies

ELLs often struggle in four key areas of GED mathematics preparation: Knowledge, Problem Solving, Vocabulary, and Math Application. *Bridging* focuses on instructional strategies in each of these areas that help ELLs access GED preparation content.

➢ Bridging Knowledge

While the basic mechanics of mathematics are standard from one culture to the next, ELLs can often lack cultural and content-related background knowledge that many native-born GED students have, which could slow them down while trying to answer problems in unfamiliar contexts (think understanding word problems, or problems based on standard measurement). Providing students with resources to bridge this gap is essential for their success with the GED prep materials. Here are the instructional strategies to focus on when bridging the knowledge gap.

Strategy 1: Develop and connect background knowledge, skills, and conceptual understanding to new knowledge.

Strategy 2: Use guiding questions to make connections beyond the lesson to broader math applications.

Strategy 3: Use problem-solving strategies to develop, monitor, and synthesize conceptual understanding and fluency. *(See Bridging Problem Solving)*

Strategy 4: Extend problem-solving skills and mathematical reasoning to broader math applications in life and work. *(See Bridging Math Application)*

1. *Bridging* provides references to the New Readers Press pre-HSE *Core Skills* books, *Pre-HSE Workbooks*, and *Scoreboost* in order to bridge students' background knowledge to the New Readers Press GED preparation materials. Here is an example:

Unit 2 **The Basics of Algebra, Section 1**	**Part 1** Expressions and Variables	**Part 2** Evaluating Expressions	**Part 3** Simplifying Expressions	**Part 4** Negative Exponents	**Part 5** Simplifying Radicals
Core Skills in Mathematics	Unit 4, Lesson 1: Evaluating Expressions (p. 60)				
Scoreboost Mathematics: Algebraic Reasoning	Evaluate Expressions (p. 10)			Find Powers and Roots (p. 8)	
Pre-HSE Workbook: Math 2	Using Algebraic Expressions and Variables (p. 16)				

2. Each *Bridging* lesson references the introduction of each *Math Sense* unit. These intros provide context examples that to apply the target math concepts as well as questions that connect to students' prior knowledge and experiences, providing a bridge to broader math applications in life and work.
3. Each *Bridging* lesson also references the specialized *Math Sense* Problem Solver, Tools, and Test Taker lessons to build student knowledge in these areas. When these lessons align with GED assessment targets, the GED target code is included. An example of how this is presented is shown using the example that supports Unit 2, Section 2 of *Math Sense 2: Focus on Problem Solving*.

Unit 2, Section 2	**Content (GED Target)**	**Page**
Problem Solver	Reading and Writing Equations (A.1.c, A.1.g, A.1.j)	58
Tools	Graphing Inequalities (A.3.b)	70
Test Taker	Try the Answer Choices	74

➢ Bridging Problem Solving

Conceptual understanding of mathematics is a key to successful problem solving and application. However, compared to many native-born GED students, ELLs often have had less exposure to mathematics presented in contextualized ways. Rote learning is the goal of many foreign educational systems and as a result our ELLs often lack conceptual knowledge of the math skills they have learned. Furthermore, the amount of mathematical terminology is overwhelming for ELLs. Problem solving strategies that reinforce this terminology help ELLs develop greater understanding and facility with the language of math. Here are four instructional strategies that help bridge this gap in conceptual understanding and language development.

Strategy 1:* *Preview the problem to determine problem-solving strategies and tools and predict general solutions.

Strategy 2:* *Develop conceptual understanding of mathematical problems using visual representations, think-alouds, and collaboration.

Strategy 3:* *Overcome barriers to problem solving using math models, language and structural analysis, and resources.

Strategy 4:* *Demonstrate and defend problem solving and mathematical reasoning through reverse problem solving, mental mathematics, visual representations, and peer discussions.

1. Each *Bridging* lesson provides a chart that breaks down the features of each mathematical concept such as the symbols, language, problem structure, and strategies that may be used to complete a particular math task. Although the completed chart is available (in the kind of style a student might complete it), the idea behind it is that teachers present it to students to fill out as a graphic organizer. Specifically, the graphic organizer is helpful when students are annotating and reviewing math concepts and choosing appropriate strategies to complete tasks. Here is an abbreviated example from the lesson on *Math Sense 2: Focus on Problem Solving,* Unit 2, Section 1:

	Symbols	**Language**	**Operation or Action**	**Structure**	**Strategies/Tools**
Part 1 Expressions and Variables	$x, n, +, -, =$	variable, expression, equation, inequality, product, quotient, difference, times, sum, less than, greater than, divided by, increased by, decreased by	translate, write	expressions show relationship between numbers and variables using operations; equations show what expressions equal	match operations to words
Part 2 Evaluate Expressions	$x, n, +, -, =, ()$	expression, variable, value, order of operations, negative, negation symbol, grouping symbol	evaluate, find, negate, group	do operations in grouping symbols ( )		first; negation symbol means it is opposite of the shown value	use order of operations (PEMDAS)

2. Building mathematical reasoning requires extending exploration with math concepts in concrete ways. Bridging presents sample exploratory activities that allow students to explore the concepts that will be presented in the lesson. Here is an example that supports *Math Sense 2: Focus on Problem Solving,* Unit 2, Section 1:

Parts 1–3: Evaluating Expressions	Divide students into groups of three or four. Give each group a different simple algebraic expression (for example, $a + 2b$). Tell them what the variables represent (for example, weight). Students can then choose what they want the coefficients to represent (people, apples, elephants, etc.). Students can then connect the expression to real life by drawing the coefficients within the expression: $a +$  b. Next, they replace variables a and b with values of their choosing: (1,000 lbs)  + (3,000 lbs). Finally, students can find the value of the expression (in this case, 7,000 lbs). Repeat this several times with different algebraic expressions that allow them to explore different operations and scenarios.
Parts 4–5: Using Rules of Exponents and Square Roots	One great way to get a job is through networking. Have students discuss networking through social media. How does it work? How can someone reach people they don't even know? Tell students that they are (hypothetically) going to use social media to find a job. On Day 1, they ask their 3 closest friends if they know of any job openings. On Day 2, those friends ask 3 of their friends. On Day 3, each of those friends ask 3 more friends and so on until a week has passed. In one week, how many people will be helping you find a job? What if you asked 5 friends and they each asked 5 and so on for one week?

3. Each lesson also offers suggestions on how to support ELLs in that specific lesson. This might include using math models as references, language and structural analysis to interpret problems and tasks, and providing resources such as peers, references, and vocabulary-building materials.

4. Perhaps the greatest gift we can give our ELLs in the math classroom is the opportunity to discuss their mathematical reasoning and methods. This offers the benefits of enhancing an individual's own mathematical reasoning and skills, helping others enhance their mathematical reasoning and skills, holding students accountable for their learning, and allowing students to try using and enhancing their understanding of mathematical terminology. *Bridging* provides examples for how to create these opportunities within each lesson.

➢ Bridging Vocabulary

Vocabulary building for ELLs is critical for success in GED preparation. *Bridging* focuses on strategies for building mathematical content vocabulary. Content vocabulary refers to subject-related words that are often exclusive to that subject area or have a particular meaning within that subject area. Math terminology is both extensive and precise. *Bridging* describes strategies for both explicit practice with math content vocabulary and contextualized practice through discussion activities. Here are some instructional strategies for providing the kinds of exposure and practice ELLs need to add new words to their productive vocabulary.

Strategy 1: Identify the component parts and usage of new words to interpret their meanings.

Strategy 2: Use context clues to interpret new words.

Strategy 3: Utilize vocabulary-building resources.

Strategy 4: Build a deeper knowledge of words through math application tasks and collaborative discussions.

Strategy 5: Memorize words through repetitive study such as using flashcards (digital or print) and notes.

Bridging Appendix B contains lists of mathematical content words for each regular lesson of the *Math Sense* texts. Each math lesson contains an alphabetized list of terminology for each part of each lesson. Appendix C provides a template for students to capture the types of practices suggested in the Bridging Vocabulary strategies. Students will fill in the template with information for the target words (teacher's choice) of each lesson as you lead them through the following strategies:

1. First, present the shortest form of the word (the base word, often the verb form), followed by other commonly used word forms (if available). Examine prefixes and suffixes and their impact on word meaning and usage.
2. Read the word as used in the context of the text and discuss possible meanings given context clues and word form.
3. Have students find (electronically or in print) the definition or translation of the base form and, if different, the form used in context and note these definitions in the space provided for future reference and study.
4. Gradually build a deeper knowledge of the word by having students use the word in a sentence frame, guided discussion, and an original sentence.
5. The high volume of mathematical terminology requires repeat exposure to the words over time. Word walls, intentionally including the words in questions to students and when eliciting responses from them, and explicit reminders to use the vocabulary in verbal tasks provide built-in reinforcement. However, this is often not enough so it is important that students learn ways to study words independently. Flashcards or websites that offer repetitive vocabulary practice are excellent ways for students to do this. Students may also use their notes, however, they will need to do repetitive activities, similar to flashcard practice, and not simply read and reread their notes.

Example:

Word or ***phrase*** (usage): symbol (if applicable)	estimate (verb)(noun)
Definition/Translation:	*(Student generated)*
In Context:	"An **estimate** can also help you see whether your answer makes sense."
Sentence Frame:	*I **estimate** the cost of the house will exceed ____________.*
Guided Discussion:	*Do you use **estimation** at the grocery store?*
Original Math Sentence:	(Student generated)

In addition to the content and academic words targeted in the lessons, *Bridging* contains a list of test-taking words in Appendix A, p. 160, that show up in the GED test and prep materials. It is crucial that students have a working knowledge of these words. Providing a reference list for students to study and refer to will greatly enhance their ability to participate in GED prep practices and activities.

➢ Bridging Math Application

This section attempts to answer the age-old question, "Why do I need to know this anyway?" Let's be honest, some mathematics may only be used in professions that require advanced mathematics. However, do we, or our students, know what professions they may one day end up in? What about that student who came to you with fantastic math skills but very little conceptual knowledge and limited English proficiency? Will building their conceptual understanding and language proficiency lead to one of these careers? Anything is possible in the world of teaching and learning, as any teacher can attest to, yet this line of thought only gets us so far. *Bridging* attempts to provide activities that allow students to practice math skills and concepts in real-life situations in ways that allow students to show their synthesis of learning and ability to apply their learning to real-life contexts. The following instructional strategies can be employed with this goal in mind:

Strategy 1: ***Prepare for math applications by identifying the problem type and the problem-solving strategies and tools.***

Strategy 2: ***Organize the problem using visual, symbolic, and written representations.***

Strategy 3: ***Overcome barriers to problem solving using math models, language and structural analysis, and resources.***

Strategy 4: ***Demonstrate and defend problem-solving application and mathematical reasoning through reverse problem solving, mental mathematics, visual representations, and peer discussions.***

1. Each *Bridging* lesson suggests that students orient themselves to the problem type and the problem-solving strategies and tools they may utilize. In other words, students must develop strategies that allow them to dissect math problems to determine what they need to do and how they need to do it. *Bridging* provides a graphic organizer to help with this dissection.
2. Each lesson encourages students to develop strategies for task planning, implementation, and completion. Specifically, each lesson suggests that students create their own graphic organizer that outlines the steps they need to take and gives them a place to record their answers.
3. Similar to Bridging Problem Solving Strategy 3, each Bridging Math Application section suggests ways to support ELLs in each lesson. These supports may be examples of using math models as references, language and structural analysis to interpret problems and tasks, and resources such as peers, references, and vocabulary-building resources.
4. This strategy offers another built-in opportunity to allow students to practice their language skills by discussing their reasoning and mathematical methods. Each Bridging Math Application section provides examples of how to structure activities that require students to defend their mathematical reasoning and methods and probe those of others.

Unit 1

WHOLE NUMBERS

Skills-Based Questions:

1. How do you combine numbers to find a total amount? Why would you do this? *(Part 1)*
2. How do you find the difference between numbers? Why would you do this? *(Part 2)*
3. What is a quick way to add the same number (or amount) over and over? Why would you do this? *(Part 3)*
4. How do you separate a number (or amount) into equal parts? Why would you do this? *(Part 4)*

Math Sense 1: Focus on Operations: Part 1, p. 16; Part 2, p. 18; Part 3, p. 20; Part 4, p. 22

	Learning Goals:	GED
Knowledge Goals:	1. Describe the operation of addition and in which situations it is used. *(Part 1)*	Q.2.a
	2. Describe the operation of subtraction and in which situations it is used. *(Part 2)*	
	3. Describe the operation of multiplication and in which situations it is used. *(Part 3)*	
	4. Describe the operation of division and in which situations it is used. *(Part 4)*	
Problem-Solving Goals:	1. Use addition, subtraction, multiplication, and division of whole numbers to solve problems. *(Parts 1–4)*	Q.2.a
	2. Solve one-step mathematical problems using the four operations with whole numbers. *(Parts 1–4)*	Q.2.e
	3. Estimate to predict and check solutions that require exact answers and solve problems for which an estimate is an appropriate solution. *(Problem Solver)*	
Vocabulary Goals:	1. Define key mathematical terms.	
	2. Determine the meaning of unknown vocabulary using context clues, word forms, and parts of speech.	
	3. Apply new vocabulary to mathematical tasks and discussions.	
Math Application Goals:	1. Apply the four operations to solve real-life math problems. *(Parts 1–4)*	Q.2.a
	2. Defend math applications and reasoning to others. *(Parts 1–4)*	Q.2.e

Sample Instructional Support Strategies

➢ Bridging Knowledge

Strategy 1: ***Develop and connect background knowledge, skills, and conceptual understanding to new knowledge.***

Strategy 2: ***Use guiding questions to make connections beyond the lesson to broader math applications.***

Strategy 3: ***Use problem-solving strategies to develop, monitor, and synthesize conceptual understanding and fluency.*** *(See also Bridging Problem Solving)*

Strategy 4: ***Extend problem-solving skills and mathematical reasoning to broader math applications in life and work.*** *(See Bridging Math Application)*

1. Evaluate students' knowledge of the following mathematical skills. Utilize the chart below to develop student content knowledge as necessary.

Unit 1 **Whole Numbers**	**Part 1** Addition	**Part 2** Subtraction	**Part 3** Multiplication	**Part 4** Division
Core Skills in Mathematics	Unit 1, Lesson 1: Number Line (p. 12); Lesson 2: Understanding Place Value (p. 15); Lesson 3: Rounding Numbers and Estimating (p. 19); Lesson 4: Performing Operations on Whole Numbers and Decimals (p. 23)		Unit 1, Lesson 3: Rounding Numbers and Estimating (p. 19); Lesson 4: Performing Operations on Whole Numbers (p. 23); Lesson 5: Finding Common Factors and Multiples (p. 27)	
Scoreboost Mathematics: Fractions, Decimals, Percents, and Proportions	Use a Problem-Solving Plan (p. 4); Estimate to Solve Problems (p. 6); Apply Number Properties to Problem Solving (p. 8)			
Pre-HSE Workbook: Math 1	Representing Numbers on a Number Line (p. 8); Rounding Numbers and Estimating (p. 10); Adding and Subtracting Whole Numbers and Decimals (p. 12)		Rounding Numbers and Estimating (p. 10); Multiplying and Dividing Whole Numbers and Decimals (p. 14)	

2. The unit preview for each unit in *Math Sense 1: Focus on Operations* provides a list of context examples in which to apply the target math concepts as well as questions that connect to students' prior knowledge and experience (for Unit 1, p. 15). These questions also provide a bridge to broader math applications in life and work.

3. Each unit also provides specialized lessons that focus on problem solving, using tools, and test taking techniques using the math skills taught in the lesson. Utilize these lessons to build student knowledge in these areas.

Unit 1	**Content (GED Target)**	**Page**
Problem Solver	Mental Math and Estimation (Q.2.a) The Five-Step Plan (word problems)	26 28
Tools	Using Your Calculator	30
Test Taker	Choose an Operation	32

➤ Bridging Problem Solving

Strategy 1: ***Preview the problem to determine problem-solving strategies and tools and predict general solutions.***

Strategy 2: ***Develop conceptual understanding of mathematical problems using visual representations, think-alouds, and collaboration.***

Strategy 3: ***Overcome barriers to problem solving using math models, language and structural analysis, and resources.***

Strategy 4: ***Demonstrate and defend problem solving and mathematical reasoning through reverse problem solving, mental mathematics, visual representations, and peer discussions.***

1. For Parts 1–4 of Unit 1, orient students to features of each mathematical problem or problem set such as the symbols, language, and structure of the problem. Students should identify the math language and symbols to determine the math operation to be used, and use the math operation and problem structure to determine which strategies to use for finding a solution. Although it is not necessary for students to fill out a graphic organizer for each math problem they attempt, completing the following graphic organizer is helpful for annotating and reviewing math concepts and choosing appropriate strategies to complete tasks. The following is an example of how a student (with guidance) might fill out this graphic organizer:

	Symbols	Language	Operation	Structure	Strategies/Tools
Part 1 Addition	+	how much/many; total	add/addition	place values lined up	line up place values; estimate first; use regrouping
Part 2 Subtraction	–	difference between; how much more; how far apart	subtract/ subtraction	place values lined up	line up place values; estimate first; use regrouping
Part 3 Multiplication	×	in a year; in 4 months	multiply/ multiplication	line up right side	estimate; times tables; use regrouping
Part 4 Division	÷	per week; in 1 month; length of each	divide/ division	dividend inside bracket/divisor out	estimate using compatible numbers; use four-step process

Students can use estimation to predict a general solution to a problem against which they can check their final answer. This strategy is suggested throughout the *Math Sense 1: Focus on Operations* text.

2. For whole number numeric equations using the four operations, it is not likely that students studying at the pre-GED level will need to develop conceptual understanding of them. If they do, however, providing a real life context for the equation and asking students to visually represent it will help build this conceptual understanding. For example, if the equation is 38 – 16, a real life problem could be the following:

> There is a total of 38 math textbooks in the classroom and 16 of them are being used. Another teacher asks to borrow the leftover textbooks for her 19 students. How many textbooks are available? Are there enough?

Students can depict this with pictures or items, then use the visual to develop a conceptual understanding of subtraction.

3. More likely, students may need help with the math skills used in the operations and not the overall conceptual understanding. In that case, a cheat sheet such as the graphic organizer above can remind them of how to attack a certain operation. Furthermore, analyzing the structure of the problem will help them determine what strategies to employ. For example, given a word problem, students will need to analyze the language to both understand what they are being asked to solve and to find the facts they need to solve it. *Math Sense 1: Focus on Operations* provides a short list of signal words for math operations. A longer list is provided on p. 159 of this book.

4. Perhaps the strongest way to evaluate and promote students' conceptual understanding and math skill fluency is to have them defend their mathematical reasoning. A rubric for self, peer, or teacher evaluation can guide this exchange and help generate constructive feedback to promote language development, skill fluency, conceptual understanding, and problem-solving strategies. This may be as simple as a "turn and talk" in which each student turns to a partner and explains how he/she solved a problem or as complex as a presentation that requires verbal explanation, visual depiction, and skill demonstration. Regardless of the level of the task chosen, it is important that students be held accountable for using precise mathematical language and supporting their explanations with sound mathematical reasoning.

The student:	Poorly (1 point)	Somewhat Well (2 points)	Well (3 points)
Explained the problem...	**inaccurately**, AND/OR by using the words directly from the problem.	**accurately** using own words BUT used only **1 or 2** appropriate math terms.	**accurately** using own words AND **3 or more** appropriate math terms.
Demonstrated how to solve the problem...	by showing **inaccurate** skills AND/OR explaining it **without** using math terms.	by showing **accurate** skills BUT explained using only **1 or 2** appropriate math terms.	by showing **accurate** skills AND explained using **3 or more** appropriate math terms.
Explained why it was solved that way...	**without support** OR with **inaccurate support** from the problem and/or math knowledge.	**with accurate support** from the problem OR math knowledge.	**with accurate support** from the problem AND math knowledge.
Described how the solution was checked...	**without** an appropriate math strategy.	using **1 appropriate** math strategy.	using **2 appropriate** math strategies.

➢ Bridging Vocabulary

Strategy 1: Identify the component parts and usage of new words to interpret their meanings.

Strategy 2: Use context clues to interpret new words.

Strategy 3: Utilize vocabulary-building resources.

Strategy 4: Build a deeper knowledge of words through math application tasks and collaborative discussions.

Strategy 5: Memorize words through repetitive study such as using flashcards (digital or print) and notes.

1. First, present the shortest form of the word, referred to in this text as the "base word" in the case of academic words and some subject-specific terms. Follow the base form with other commonly used word forms (if available). Examine prefixes and suffixes and their impact on word meaning and part of speech.
2. Read the word as used in the context of the text and discuss possible meanings given context clues and word form.
3. Have students find (electronically or in print) the definition or translation of the base form and, if different, the form used in context and note these definitions in the space provided for future reference and study.
4. Gradually build a deeper knowledge of the word by having students use the word in a sentence frame, guided discussion, and an original sentence within a mathematical context.

Sentence Frame:	*The **sum** of ______________ and ______________ is ______________.*
Guided Discussion:	*How do you calculate the **sum**? Give an example of a situation in which you calculate the sum.*
Original Math Sentence:	______________________________

Encourage students to use these words in math applications and collaborative discussions such as the task described in Bridging Problem Solving, Strategy 4.

5. The high volume of mathematical terminology requires repeat exposure to the words over time. Word walls, intentionally including the words in questions to students and when eliciting responses from them, and explicit reminders to use the vocabulary in verbal tasks provide built-in reinforcement. However, this is often not enough, so it is important that students learn ways to study words independently. Flashcards or websites that offer repetitive vocabulary practice are excellent ways for students to do this. Students may also use their notes, however, they will need to do repetitive activities, similar to flashcard practice, and not simply read and reread their notes.

➤ Bridging Math Application

Strategy 1: ***Prepare for math applications by identifying the problem type and the problem-solving strategies and tools.***

Strategy 2: ***Organize the problem using visual, symbolic, and written representations.***

Strategy 3: ***Overcome barriers to problem solving using math models, language and structural analysis, and resources.***

Strategy 4: ***Demonstrate and defend problem-solving application and mathematical reasoning through reverse problem solving, mental mathematics, visual representations, and peer discussions.***

1. Each part of this lesson lends itself to a variety of math application tasks that allow students to synthesize, apply, or extend their mathematical knowledge and skills. Whichever math application task you choose, be sure to orient students to the problem type and the problem-solving strategies and tools they may utilize. The following activity is an example of student directions for a math application task that synthesizes all four operations.

In teams, have students collect information from classmates about how many shirts they or their households have. Use this information to find the total number of shirts the group has as a whole, the average number (divide the total by the number of people in the group) of shirts each person has, and how you compare to the average. Finally, use the average to predict the total number of shirts all the students in the school have combined. This activity can be done counting any item, for example, cell phones, credit cards, children, etc.

Before engaging in problem solving, have students analyze the directions to determine what is being asked of them, and use this information to determine which strategies to use to complete the task. The following chart begins this process to provide a glimpse of how this might look.

Directions	**Operation or Action** *Signal Words*	**Strategies and Tools**
In teams, collect information from classmates, about how many shirts they or their households have. Use this information to find the total number of shirts the group has as a whole …	Addition *how many, total number*	Line up place values of all numbers. Estimate total. (Use 5s and 10s) Add up and regroup to next place value as necessary. Compare the actual total to your estimate to see if the calculation is reasonable.
the average number of shirts each person has … (divide the total by the number of people in the group)		

2. Have students meet in teams and choose an item they want to count. They will need to describe the problem and how to handle each component. Have them create a checklist or other graphic organizer listing the steps they need to complete to accomplish the task.

- ☑ Choose an item.
- ☑ Write a survey question.
- ☑ Collect data.
- ☑ Add total.
- ☑ Divide total by total number of students surveyed.
- ☑ Subtract to compare my amount. (Do I have more or less?)
- ☑ Estimate the number of students in the school.
- ☑ Multiply the total number of students by the average.

3. One resource that is often overlooked is peer work. Working with peers is a good way for students to develop and hone their mathematical knowledge and skills through discussions and tasks that require them to explain (and sometimes defend) their own mathematical reasoning and probe that of others. Using the example in Bridging Math Application, students worked in small groups to organize and accomplish the task. During collaborative work, students can practice both precise mathematical terms to describe their ideas and mathematical reasoning as well as the discourse prompts (p. 159) necessary to work well in a group.

Polite disagreement: *I see what you're saying but I think* ____________. *Another way to think about this is* ____________.

Encourage participation: *What do you think? I'd like to hear what you have to say about* ____________.

Probe others' ideas: *Could you explain that further? What did you mean by* ____________*?*

Paraphrase: *So, do we agree that* ____________*? I think you're saying that* ____________.

Present examples: *An example of this is* ____________. ____________ *is an example of* ____________.

4. As mentioned in Strategy 3, collaboration with students, including student discussions, helps students overcome barriers to problem solving but it also makes them accountable for their own mathematical skills and reasoning. This can be accomplished through a jigsaw activity. Have each group pair up with another group. Each group takes a turn describing to the other group their survey and the results of their calculations. Stress the importance of using precise mathematical language in their explanations and supporting them with sound mathematical reasoning. A good way to set up this exchange is to provide a checklist of student expectations.

- ☑ Introduce your survey question.
- ☑ Take turns with your group members in describing each step of the task.
- ☑ Take turns explaining the mathematical reasoning behind each of the calculations.
- ☑ Use precise mathematical language.

➤ Assessment & Next Steps

Students should complete the practice activities included in each Math Sense 1: Focus on Operations lesson. Evaluate which learning goals were not met and remediate by using other resources, such as those identified in the Bridging Knowledge section. Upon successful completion, continue to the next unit.

Unit 2

DECIMALS: Section 1

Skills-Based Questions:

1. When is it necessary to use a number with decimal points? What information does the decimal point tell us about the numbers to the left and to the right of it? *(Part 1)*
2. How do you represent decimals in numbers and words? *(Part 2)*
3. What is rounding? How do you round decimals to a whole number or a particular place value? *(Part 3)*
4. How do you know which decimal is greater or less than another? *(Part 4)*

Math Sense 1: Focus on Operations: Part 1, p. 38; Part 2, p. 40; Part 3, p. 42; Part 4, p. 44

	Learning Goals:	**GED**
Knowledge Goals:	1. Describe what a decimal point means and the relationship between the numbers to the left and right of it. *(Part 1)*	Q.1.a
	2. Describe the place values to the left and right of a decimal point. *(Part 2)*	
	3. Describe the rounding process when rounding decimals to whole numbers or to particular place values. *(Part 3)*	
	4. Describe the relationship (greater than, less than, equal to) between different decimals. *(Part 4)*	
Problem-Solving Goals:	1. Order decimals, including on a number line. *(Part 1)*	Q.1.a
	2. Identify place values to the left and write of the decimal point and write decimals in words and numbers *(Part 2)*	
	3. Round decimals to whole numbers and named place values. *(Part 3)*	
	4. Compare decimals to determine which is greater or less than. *(Part 4)*	
Vocabulary Goals:	1. Define key mathematical terms.	
	2. Determine the meaning of unknown vocabulary using context clues, word forms, and parts of speech.	
	3. Apply new vocabulary to mathematical tasks and discussions.	
Math Application Goals:	1. Apply conceptual understanding of decimals to real life situations. *(Parts 1–4)*	Q.1.a
	2. Defend math applications and reasoning to others. *(Parts 1–4)*	

Sample Instructional Support Strategies

➢ Bridging Knowledge

Strategy 1: ***Develop and connect background knowledge, skills, and conceptual understanding to new knowledge.***

Strategy 2: ***Use guiding questions to make connections beyond the lesson to broader math applications.***

Strategy 3: ***Use problem-solving strategies to develop, monitor, and synthesize conceptual understanding and fluency.*** *(See also Bridging Problem Solving)*

Strategy 4: ***Extend problem-solving skills and mathematical reasoning to broader math applications in life and work.*** *(See Bridging Math Application)*

1. Evaluate students' knowledge of the following mathematical skills. Utilize the chart below to develop student content knowledge as necessary.

Unit 2 **Decimals, Section 1**	**Part 1** Understanding Decimals	**Part 2** Writing Decimals	**Part 3** Rounding Decimals	**Part 4** Comparing Decimals
Core Skills in Mathematics	Unit 1, Lesson 1: Representing Numbers on a Number Line (p. 12); Lesson 2: Understanding Place Value (p. 15)		Unit 1, Lesson 3: Rounding Numbers and Estimating (p. 19)	Unit 1, Lesson 1: Representing Numbers on a Number Line (p. 12); Lesson 2: Understanding Place Value (p. 15)
Pre-HSE Workbook: Math 1	Representing Numbers on a Number Line (p. 8)		Rounding Numbers and Estimating (p. 10)	

2. The unit preview for each unit in *Math Sense 1: Focus on Operations* provides a list of context examples in which to apply the target math concepts as well as questions that connect to students' prior knowledge and experience (for Unit 2, p. 37). These questions also provide a bridge to broader math applications in life and work.

3. Each unit also provides specialized lessons that focus on problem solving, using tools, and test taking techniques using the math skills taught in the lesson. Utilize these lessons to build student knowledge in these areas.

Unit 2, Section 1	Content	Page
Problem Solver	Solving Multistep Problems	48
	Powers of Ten	56
	Figuring Unit Price and Total Cost	58
Tools	Calculators and Decimals	60
Test Taker	Use Estimation to Choose an Answer	62

➢ Bridging Problem Solving

Strategy 1: ***Preview the problem to determine problem-solving strategies and tools and predict general solutions.***

Strategy 2: ***Develop conceptual understanding of mathematical problems using visual representations, think-alouds, and collaboration.***

Strategy 3: ***Overcome barriers to problem solving using math models, language and structural analysis, and resources.***

Strategy 4: ***Demonstrate and defend problem solving and mathematical reasoning through reverse problem solving, mental mathematics, visual representations, and peer discussions.***

1. For Parts 1–4 of Unit 2, orient students to features of each mathematical concept such as the symbols, language, and structure. Students should identify the math language and symbols to determine what is being asked of them and use this information to determine which strategies to use to complete the task. Although it is not necessary for students to fill out a graphic organizer for each math problem they attempt, completing the following graphic organizer is helpful for annotating and reviewing math concepts and choosing appropriate strategies to complete tasks. The following is an example of how a student (with guidance) might fill out this graphic organizer:

	Symbols	Language	Operation or Action	Structure	Strategies/Tools
Part 1 Understanding Decimals	.	place value, whole numbers, tens/ten*ths*	understand	whole numbers = 1 & up; numbers to right of (.) = less than 1	use a number line to order decimals; line up place values in columns to left and right of (.)
Part 2 Writing Decimals	.	*and* = decimal point, *ths* for places right of decimal point	write	place values increase moving left, decrease moving right	use leading zero and placeholder zero; write spaces as placeholders before and after decimal point
Part 3 Rounding Decimals	.	rounding, whole number, place value	round	use one place value to the right to round	identify which place value to round to; use 1 number to the right; 0–4 rounds down and 5–9 rounds up
Part 4 Comparing Decimals	< > + =	greater than, less than, equal to	compare	value of number increase to the left and decreases to the right	add placeholder zeros so that each number has the same number of place values, then compare like whole numbers

2. Providing a real-life context for a math concept is an important way to build conceptual understanding. Situations involving money are often used to help students understand decimals. Explain that for money, dollars are the units with $1 being the lowest whole number. Anything less than $1 must be represented with decimals. Provide a variety of examples of representing different change as decimals, focusing on quarters to give students practice with .25, .5, and .75. With this baseline, have students form teams, choose their own unit, and establish their own lowest whole number. Students should then provide three examples of what decimals would look like (using a drawing or the actual item) using that unit and write the corresponding number. Students can also use their example to compare decimals.

Unit = Donuts

The whole = 1 donut	**Example 1**	**Example 2**	**Example 3**
Drawing:			
Number/decimal: 1	0.5	0.25	0.75

3. *Math Sense 1: Focus on Operations* provides a Decimal, Fraction, and Percent Equivalencies chart (p. 205) that contains the most common decimals and their fraction and percent equivalencies. Students can use this resource to remind them how to represent these equivalencies in the various mathematical ways. Displaying math models of these equivalencies is also helpful. These can be student generated. Divide students into pairs. Give each pair a blank poster and a common decimal from this reference sheet. Challenge students to depict the decimal on the poster in as many ways as they can think of in a 10-minute period. Provide miscellaneous items (glue or tape) as well as markers for drawing and labeling.

Example: Poster for 0.75

The poster shows an illustrated ¾ cup, a popsicle stick cut to a ¾ length and glued down, three quarters taped down, and a magazine photo of a football field cut at 75 yards. The title of the poster is ***What does 0.75 look like?*** and each example shown is labeled with the unit and *the whole* it represents. For example, the photo of the football field is labeled with: "Units = football fields; Whole = 1 football field." The poster is then displayed in the classroom along with others.

4. Perhaps the strongest way to evaluate and promote students' conceptual understanding is to have them defend their mathematical reasoning. The poster example above provides an excellent way to evaluate students understanding. To take this up a notch, have students explain the choices they made for their poster and why. This can be accomplished through a jigsaw activity. Have pairs of students team up with one or two other pairs. Each pair takes a turn explaining their poster, their examples, and the mathematical reasons for their choices. Stress the importance of using precise mathematical language in their explanations and supporting them with sound mathematical reasoning. A good way to set up this exchange is to provide a checklist of student expectations.

☑ Introduce your poster with the title.

☑ Take turns with your partner describing each of the examples.

☑ Use precise mathematical language, including saying your decimal correctly.

☑ Explain how your example mathematically represents the decimal (your mathematical reasoning).

➤ Bridging Vocabulary

Strategy 1: ***Identify the component parts and usage of new words to interpret their meanings.***

Strategy 2: ***Use context clues to interpret new words.***

Strategy 3: ***Utilize vocabulary-building resources.***

Strategy 4: ***Build a deeper knowledge of words through math application tasks and collaborative discussions.***

Strategy 5: ***Memorize words through repetitive study such as using flashcards (digital or print) and notes.***

1. First, present the shortest form of the word, referred to in this text as the "base word" in the case of academic words and some subject-specific terms. Follow the base form with other commonly used word forms (if available). Examine prefixes and suffixes and their impact on word meaning and part of speech.
2. Read the word as used in the context of the text and discuss possible meanings given context clues and word form.
3. Have students find (electronically or in print) the definition or translation of the base form and, if different, the form used in context and note these definitions in the space provided for future reference and study.
4. Gradually build a deeper knowledge of the word by having students use the word in a sentence frame, guided discussion, and an original sentence within a mathematical context.

Sentence Frame:	*The **decimal point** separates the* ________________ *from the* ________________.
Guided Discussion:	*Why do we need a **decimal point** in some numbers? Give an example to support your answer.*
Original Math Sentence:	________________________________

Encourage students to use these words in math applications and collaborative discussions such as the task described in Bridging Problem Solving Strategy 4.

5. The high volume of mathematical terminology requires repeat exposure to the words over time. Word walls, intentionally including the words in questions to students and when eliciting responses from them, and explicit reminders to use the vocabulary in verbal tasks provide built-in reinforcement. However, this is often not enough so it is important that students learn ways to study words independently. Flashcards or websites that offer repetitive vocabulary practice are excellent ways for students to do this. Students may also use their notes, however, they will need to do repetitive activities, similar to flashcard practice, and not simply read and reread their notes.

➤ Bridging Math Application

Strategy 1: ***Prepare for math applications by identifying the problem type and the problem-solving strategies and tools.***

Strategy 2: ***Organize the problem using visual, symbolic, and written representations.***

Strategy 3: ***Overcome barriers to problem solving using math models, language and structural analysis, and resources.***

Strategy 4: ***Demonstrate and defend problem-solving application and mathematical reasoning through reverse problem solving, mental mathematics, visual representations, and peer discussions.***

1. Parts 1–4 of this lesson each lend themselves to a variety of math application tasks that allow students to synthesize, apply, or extend their mathematical knowledge and skills. Whichever math application task you choose, be sure to orient students to the problem type and the problem-solving strategies and tools they may utilize. The following activity, synthesizes all four operations into one real-life application.

You want to get a credit card with the lowest interest rate. Research five different credit cards online. Compare their interest rates and list them from lowest to highest. Identify the lowest rate and explain how you know that it is lower than the other rates.

	Sample Mathematical Application	**Strategies and Tools**
Part 1:	Find and note the interest rate for five different credit cards.	Identify where the decimal point is and copy the numbers correctly on either side.
Part 2:	Read the interest rates.	Say each interest rate out loud to practice the place values.
Part 4: (before Part 3)	Put the interest rates in order.	Line up the decimal points and stack the interest rates. Plug in zero placeholders as needed.
Part 3: (after Part 4)	Check the order of your list by rounding to the nearest whole number.	Look at the place value to the right of the decimal point to round (0–4 rounds down; 5–9 rounds up). Compare the order to rounded values to see if the order is reasonable.

2. This example is simple as it deals with understanding decimals and not manipulating them. As such, listing the interest rates in order does organize the information for future manipulation.
3. Students can use the tools and strategies they have learned leading up to this task, like the *Math Sense 1: Focus on Operations* visuals of place values and the Decimal, Fraction, and Percent Equivalencies chart (p. 205). Furthermore, the posters they created to model common fractions may help them compare fractions they find that represent the credit card interest rates.
4. The math application task requires that students explain how they know that the interest rate they chose is lower than the other rates. Students must defend their choice by explaining their mathematical reasoning behind it. Provide a checklist to let students know what is expected during their explanation.

☑ Identify the credit card you chose and read its interest rate correctly.

☑ Describe how you know the rate you chose is less than the other rates (your mathematical reasoning).

☑ Use precise mathematical language, including saying the decimals correctly.

➢ Assessment & Next Steps

Students should complete the practice activities included in each *Math Sense 1: Focus on Operations* lesson. Evaluate which learning goals were not met and remediate by using other resources, such as those identified in the Bridging Knowledge section. Upon successful completion, continue to the next section of the unit.

DECIMALS: Section 2

Skills-Based Questions:

1. How do you add and subtract decimals? In which situations do you do this? *(Part 5)*
2. How do you multiply and divide decimals? In which situations do you do this? *(Parts 6–7)*

Math Sense 1: Focus on Operations: Part 5, p. 46; Part 6, p. 52; Part 7, p. 54

	Learning Goals:	**GED**
Knowledge Goals:	1. Describe how to add and subtract decimals and in which situations this is used. *(Part 5)*	Q.2.a
	2. Describe how to multiply and divide decimals and in which situations this is used. *(Parts 6–7)*	Q.2.e
Problem-Solving Goals:	1. Use addition, subtraction, multiplication, and division of decimals to solve problems. *(Parts 5–7)*	Q.2.a Q.2.e
Vocabulary Goals:	1. Define key mathematical terms.	
	2. Determine the meaning of unknown vocabulary using context clues, word forms, and parts of speech.	
	3. Apply new vocabulary to mathematical tasks and discussions.	
Math Application Goals:	1. Apply the four operations to solve real-life math problems using decimals. *(Parts 5–7)*	Q.2.a
	2. Defend math applications and reasoning to others. *(Parts 5–7)*	Q.2.e

Sample Instructional Support Strategies

➢ Bridging Knowledge

Strategy 1: ***Develop and connect background knowledge, skills, and conceptual understanding to new knowledge.***

Strategy 2: ***Use guiding questions to make connections beyond the lesson to broader math applications.***

Strategy 3: ***Use problem-solving strategies to develop, monitor, and synthesize conceptual understanding and fluency.*** *(See also Bridging Problem Solving)*

Strategy 4: ***Extend problem-solving skills and mathematical reasoning to broader math applications in life and work.*** *(See Bridging Math Application)*

1. Evaluate students' knowledge of the following mathematical skills. Utilize the chart below to develop student content knowledge as necessary.

Unit 2 **Decimals, Section 2**	**Part 5** Adding and Subtracting Decimals	**Part 6** Multiplying Decimals	**Part 7** Dividing Decimals
Core Skills in Mathematics	Unit 1, Lesson 2: Understanding Place Value (p. 15); Lesson 4: Performing Operations on Whole Numbers and Decimals (p. 23)	Unit 1, Lesson 4: Performing Operations on Whole Numbers and Decimals (p. 23); Lesson 5: Finding Common Factors and Multiples (p. 27)	
Scoreboost Mathematics: Fractions, Decimals, Percents, and Proportions	Use a Problem-Solving Plan (p. 4); Solve Decimal Problems Using the Calculator (p. 16)		
Pre-HSE Workbook: Math 1	Adding and Subtracting Whole Numbers and Decimals (p. 12)	Multiplying and Dividing Whole Numbers and Decimals (p. 14)	

2. The unit preview for each unit in *Math Sense 1: Focus on Operations* provides a list of context examples in which to apply the target math concepts as well as questions that connect to students' prior knowledge and experience (for Unit 2, p. 37). These questions also provide a bridge to broader math applications in life and work.

3. Each unit also provides specialized lessons that focus on problem solving, using tools, and test taking techniques using the math skills taught in the lesson. Utilize these lessons to build student knowledge in these areas.

Unit 2, Section 2	Content (GED Target)	Page
Problem Solver	Solving Multistep Problems (Q.2.a, Q.2.e) Powers of Ten Figuring Unit Price and Total Cost	48 56 58
Tools	Calculators and Decimals	60
Test Taker	Use Estimation to Choose an Answer	62

➢ Bridging Problem Solving

Strategy 1: ***Preview the problem to determine problem-solving strategies and tools and predict general solutions.***

Strategy 2: ***Develop conceptual understanding of mathematical problems using visual representations, think-alouds, and collaboration.***

Strategy 3: ***Overcome barriers to problem solving using math models, language and structural analysis, and resources.***

Strategy 4: ***Demonstrate and defend problem solving and mathematical reasoning through reverse problem solving, mental mathematics, visual representations, and peer discussions.***

1. For Parts 5–7 of Unit 2, orient students to features of each mathematical problem or problem set such as the symbols, language, and structure of the problem. Students should identify the math language and symbols to determine the math operation to be used, and use the math operation and problem structure to determine which strategies to use for finding a solution. Although it is not necessary for students to fill out a graphic organizer for each math problem they attempt, completing the following graphic organizer is helpful for annotating and reviewing math concepts and choosing appropriate strategies to complete tasks. The following is an example of how a student (with guidance) might fill out this graphic organizer:

	Symbols	Language	Operation or Action	Structure	Strategies/Tools
Part 5 Add/ Subtract Decimals	+ −	how much/many; total difference between; how much more; how far apart	add subtract	place values lined up	line up decimal points and place values to the left and right; estimate first; use regrouping
Part 6 Multiply Decimals	×	in a year; in 4 months	multiply	line up right side	estimate; times tables; powers of ten; use regrouping
Part 7 Divide Decimals	÷	per week; in 1 month; length of each	divide	dividend inside bracket/ divisor outside bracket and move decimal to right	estimate using compatible numbers; move decimal of divisor; use four-step system; powers of ten

Students can use estimation to predict a general solution to a problem against which they can check their final answer. This strategy is suggested throughout *Math Sense 1: Focus on Operations*.

2. In section 1 of *Bridging* Unit 2: Decimals, students developed a conceptual understanding of decimals (p. 18), including creating a visual representation of a common decimal. Build on this understanding by exploring different situations in which using the four operations with decimals would occur. Again, any computations using money helps build greater conceptual understanding of decimals. In fact, the consequences for decimal errors when dealing with money are quite high and therefore provide a great example for students to understand the importance of calculating decimals correctly. Here is an activity that builds this type of awareness.

Card Game: The Decimal Is in the Details!

Materials Needed: 1 full deck of cards (no jokers)

Directions: In groups of 4 to 6, one student deals 5 cards to each student (including himself/herself). Students add up the total value of their cards (Aces = 1, Jacks = 11, Queens = 12, Kings = 13). The dealer deals each student one more card. If the card is black, students will move the decimal for their total one place to the right (to increase their total). If the card is red, they move the decimal one place to the left (to decrease their total). Each student will record their totals for that round of play and then play 4 more rounds, switching dealers each time and recording their round totals each time. At the end of Round 5, students should total their points from each round. The player with the highest total wins.

3. More likely, students may need help with the math skills used in the operations and not the overall conceptual understanding. In that case, a cheat sheet such as the graphic organizer on the previous page can remind them of how to attack a certain operation. Furthermore, analyzing the structure of the problem will help them determine what strategies to employ. For example, given a word problem, students will need to analyze the language to both understand what they are being asked to solve and to find the facts they need to solve it. *Math Sense 1: Focus on Operations* provides a short list of signal words for math operations. A longer list is provided in Appendix A of this book.

4. Reversing a math problem once the solution is found is a great way to check one's answer and develop a deeper understanding of the relationship between the operations. In reverse problem solving, students begin with their answer and use the opposite operation to arrive at the components of the problem. This is particularly powerful in word problems, as reverse problem solving not only requires a thorough understanding of the original problem but conceptual understanding of math operations. During the reverse problem-solving task, teachers and peers become aware of errors in calculations, process, and reasoning. This awareness allows the teacher to focus in on the problem area and help bridge students to greater accuracy, fluency, and conceptual understanding. Here is an example of how this may be accomplished in a lesson:

Flip It!

Intermittently during a math lesson or activity, shout out "Flip it!" (or allow the students to call this out from time to time). When students hear "Flip it!" they must finish solving the problem they are working on and then turn to a partner. Each partner will take a turn sharing their reverse problem solving aloud. Partners can help each other work through this reversal and request teacher or additional peer guidance as needed.

➤ Bridging Vocabulary

Strategy 1: ***Identify the component parts and usage of new words to interpret their meanings.***

Strategy 2: ***Use context clues to interpret new words.***

Strategy 3: ***Utilize vocabulary-building resources.***

Strategy 4: ***Build a deeper knowledge of words through math application tasks and collaborative discussions.***

Strategy 5: ***Memorize words through repetitive study such as using flashcards (digital or print) and notes.***

1. First, present the shortest form of the word, referred to in this text as the "base word" in the case of academic words and some subject-specific terms. Follow the base form with other commonly used word forms (if available). Examine prefixes and suffixes and their impact on word meaning and part of speech.
2. Read the word as used in the context of the text and discuss possible meanings given context clues and word form.
3. Have students find (electronically or in print) the definition or translation of the base form and, if different, the form used in context and note these definitions in the space provided for future reference and study.
4. Gradually build a deeper knowledge of the word by having students use the word in a sentence frame, guided discussion, and an original sentence within a mathematical context.

Sentence Frame:	*The **zero placeholder** helps keep* ____________ *when you* ____________.
Guided Discussion:	*Why do we use **zero placeholders**? Do they help you with your decimal calculations?*
Original Math Sentence:	____________________

Encourage students to use these words in math applications and collaborative discussions such as the task described in Bridging Problem Solving, Strategy 4.

5. The high volume of mathematical terminology requires repeat exposure to the words over time. Word walls, intentionally including the words in questions to students and when eliciting responses from them, and explicit reminders to use the vocabulary in verbal tasks provide built-in reinforcement. However, this is often not enough so it is important that students learn ways to study words independently. Flashcards or websites that offer repetitive vocabulary practice are excellent ways for students to do this. Students may also use their notes, however, they will need to do repetitive activities, similar to flashcard practice, and not simply read and reread their notes.

➤ Bridging Math Application

Strategy 1: ***Prepare for math applications by identifying the problem type and the problem-solving strategies and tools.***

Strategy 2: ***Organize the problem using visual, symbolic, and written representations.***

Strategy 3: ***Overcome barriers to problem solving using math models, language and structural analysis, and resources.***

Strategy 4: ***Demonstrate and defend problem-solving application and mathematical reasoning through reverse problem solving, mental mathematics, visual representations, and peer discussions.***

1. Each part of this lesson lends itself to a variety of math application tasks that allow students to synthesize, apply, or extend their mathematical knowledge and skills. Whichever math application task you choose, be sure to orient students to the problem type and the problem-solving strategies and tools they may utilize. The following activity synthesizes all four operations into one real-life application.

Have students collaborate to plan food, beverages, and supplies for a classroom party. First, determine how much money you can collect from students (contributions). Next, create a shopping list that contains the total cost for all the items that will be purchased for the party. You will need to research unit costs for each item, determine how many units will be required, calculate total cost of each item, and calculate the total amount needed to purchase all items. Make sure that you have enough contributions from the class to make the purchases. Finally, calculate the average contribution (divide the total contributions by the number of students).

Before engaging in problem solving, have students analyze the directions to determine what is being asked of them and use this information to determine which strategies to use to complete the task. The following chart begins this process to provide a glimpse of how this may look.

Sample Mathematical Application	**Operation** *Signal Words*	**Strategies and Tools**
First, determine how much money you can collect from students (contributions).	Addition *how much money*	Estimate total contributions first. Line up decimals and place values of contributions. Regroup to next place value as necessary. Compare actual amount of student contributions to estimate to see if it is reasonable.
Next, create a shopping list that contains the total cost for all the items that will be purchased for the party. You will need to research unit costs for each item, determine how many units will be required, calculate total cost of each item, and calculate the total amount needed to purchase all items.		

2. Have students meet in teams to look at each component of the problem and how to handle it. The class may determine that each team take on one part of the problem or that all the teams work out the problem and then vote on the best plan. An activity like this gives students the opportunity to take charge, so step back and let them plan. Provide a feedback loop to check student plans for feasibility. A class rubric or checklist can provide that double check. Here is an example for the above activity. (Note how the rubric takes each component and outlines three levels of performance.)

Does the plan:	**No (1 point)**	**Somewhat (2 points)**	**Yes (3 points)**
Calculate student contributions...	**inaccurately** AND/OR list **some students** and their contributions?	**accurately** and list **most students** and their contributions.	**accurately** AND list **each available student** and his/her contribution?
Calculate a shopping bill...	**inaccurately** without **actual unit costs, reasonable estimates** of unit quantity, OR cost **equal to or less than** student contributions?	**accurately** with **actual unit costs, reasonable estimates** of unit quantity, OR cost **equal to or less than** student contributions.	**accurately** with **actual unit costs, reasonable estimates** of unit quantity, AND cost **equal to or less than** student contributions?
Calculate average student contribution...	**inaccurately** AND/OR from a list of **some students** and their contributions?	**accurately** from a list of **most students** and their contributions?	**accurately** from a list of **each available student** and his/her contribution?

3. One resource that is often overlooked is peer work. Working with peers is a good way for students to develop and hone their mathematical knowledge and skills through discussions and tasks that require them to defend their mathematical reasoning and question that of others. In the Bridging Math Application example, students work in small groups to plan a class party. During this collaborative work, students can practice both precise mathematical terminology to describe and defend their mathematical reasoning as well as the discourse prompts (p. 159) necessary to work well in a group.

Polite disagreement: *I see what you're saying but I think _______________. Another way to think about this is _______________.*

Encourage participation: *What do you think? I'd like to hear what you have to say about _______________.*

Probe others' ideas: *Could you explain that further? What did you mean by _______________?*

Paraphrase: *So, do we agree that _______________? I think you're saying that _______________.*

Present examples: *An example of this is _______________. _______________ is an example of _______________.*

4. As mentioned in Strategy 3, collaboration with students, including student discussions, helps students overcome barriers to problem solving but it also makes them accountable for their own mathematical skills and reasoning. Using checklists and rubrics guide students in understanding what is required for mastery of a skill and will provide a framework for students to evaluate themselves, their peers, and their teachers. For example, student performance can be evaluated after a project like the party planning task using a rubric such as this one:

The student...	Poorly (1 point)	Somewhat Well (2 points)	Well (3 points)
Explained the problem...	**inaccurately**, AND/OR by using the words directly from the problem.	**accurately** using own words BUT used only **1 or 2** appropriate math terms.	**accurately** using own words AND **3 or more** appropriate math terms.
Demonstrated how to solve the problem...	by showing **inaccurate** skills AND/OR explaining it **without** using math terms.	by showing **accurate** skills BUT explaining using only **1 or 2** appropriate math terms.	by showing **accurate** skills AND explaining using **3 or more** appropriate math terms.
Explained why it was solved that way...	**without support** OR with **inaccurate support** from the problem and/or math knowledge.	**with accurate support** from the problem OR math knowledge.	**with accurate support** from the problem AND math knowledge.
Described how the solution was checked...	**without** an appropriate math strategy.	using **1 appropriate** math strategy.	using **2 appropriate** math strategies.

➢ Assessment & Next Steps

Students should complete the practice activities included in each *Math Sense 1: Focus on Operations* lesson. Evaluate which learning goals were not met and remediate by using other resources, such as those identified in the Bridging Knowledge section. Upon successful completion, continue to the next unit.

Unit 3

FRACTIONS: Section 1

Skills-Based Questions:

1. How are decimals and fractions alike? How are they different? In which situations is using one preferred over using the other? *(Part 1)*
2. What are the different forms of fractions? Why are there different forms? *(Part 2)*
3. What are equivalent fractions? How do you know they are equivalent? *(Part 3)*
4. How do you know which fraction is greater or less than another? *(Part 4)*
5. How do you add and subtract like fractions? *(Part 5)*

Math Sense 1: Focus on Operations: Part 1, p. 68; Part 2, p. 70; Part 3, p. 72; Part 4, p. 74; Part 5, p. 76

	Learning Goals:	GED
Knowledge Goals:	1. Describe what a fraction means and its relationship to decimals. *(Part 1)*	Q.1.a
	2. Describe the relationship between the numerator and the denominator. *(Part 2)*	Q.1.b
	3. Describe the different forms of a fraction and the relationship between them. *(Part 2)*	Q.2.a
	4. Describe equivalent fractions and how to change a fraction into higher or lower terms. *(Part 3)*	
	5. Describe the relationship (greater than, less than, equal to) between different fractions. *(Part 4)*	
	6. Describe how to add and subtract like fractions. *(Part 5)*	
Problem-Solving Goals:	1. Order fractions, including on a number line. *(Part 1)*	Q.1.a
	2. Define the relationship between fractions and decimals. *(Part 1)*	Q.1.b
	3. Identify the numerator and denominator and define their relationship. *(Part 2)*	Q.2.a
	4. Define *proper* and *improper fractions* and *mixed numbers* and explain their relationship to a whole. *(Part 2)*	
	5. Identify equivalent fractions and simplify and raise their terms. *(Part 3)*	
	6. Compare fractions to determine whether they are greater than, less than, or equal to one another. *(Part 4)*	
	7. Add and subtract like fractions. *(Part 5)*	
Vocabulary Goals:	1. Define key mathematical terms.	
	2. Determine the meaning of unknown vocabulary using context clues, word forms, and parts of speech.	
	3. Apply new vocabulary to mathematical tasks and discussions.	
Math Application Goals:	1. Apply conceptual understanding of fractions to real life situations. *(Parts 1–5)*	Q.1.a
	2. Defend math applications and reasoning to others. *(Parts 1–5)*	Q.1.b
		Q.2.a
		Q.2.e

Sample Instructional Support Strategies

➢ Bridging Knowledge

Strategy 1: ***Develop and connect background knowledge, skills, and conceptual understanding to new knowledge.***

Strategy 2: ***Use guiding questions to make connections beyond the lesson to broader math applications.***

Strategy 3: ***Use problem-solving strategies to develop, monitor, and synthesize conceptual understanding and fluency.*** *(See also Bridging Problem Solving)*

Strategy 4: ***Extend problem-solving skills and mathematical reasoning to broader math applications in life and work.*** *(See Bridging Math Application)*

1. Evaluate students' knowledge of the following mathematical skills. Utilize the chart below to develop student content knowledge as necessary.

Unit 3 **Fractions, Section 1**	**Part 1** Relating Decimals and Fractions	**Part 2** Different Forms of Fractions	**Part 3** Equivalent Fractions	**Part 4** Comparing Fractions	**Part 5** Adding and Subtracting Like Fractions
Core Skills in Mathematics	Unit 1, Lesson 6: Understanding Fractions (p. 31)				Unit 1, Lesson 7: Performing Operations on Fractions and Mixed Numbers (p. 35)
Scoreboost Mathematics: Fractions, Decimals, Percents, and Proportions					Decide Which Operation to Use with Fractions (p. 18); Fraction Problems Using the Calculator (p. 22)
Pre-HSE Workbook: Math 1	Representing Numbers on a Number Line (p. 8)				Adding and Subtracting Fractions (p. 16)

2. The unit preview for each unit in *Math Sense 1: Focus on Operations* provides a list of context examples in which to apply the target math concepts as well as questions that connect to students' prior knowledge and experience (for Unit 3, p. 67). These questions also provide a bridge to broader math applications in life and work.

3. Each unit also provides specialized lessons that focus on problem solving, using tools, and test taking techniques using the math skills taught in the lesson. Utilize these lessons to build student knowledge in these areas.

Unit 3, Section 1	**Content**	**Page**
Problem Solver	Multiples and Factors Does the Answer Make Sense?	78 94
Tools	Calculators and Fractions	96
Test Taker	Know When to Use Decimals	98

➢ Bridging Problem Solving

Strategy 1: ***Preview the problem to determine problem-solving strategies and tools and predict general solutions.***

Strategy 2: ***Develop conceptual understanding of mathematical problems using visual representations, think-alouds, and collaboration.***

Strategy 3: ***Overcome barriers to problem solving using math models, language and structural analysis, and resources.***

Strategy 4: ***Demonstrate and defend problem solving and mathematical reasoning through reverse problem solving, mental mathematics, visual representations, and peer discussions.***

1. For Parts 1–5 of Unit 3, orient students to features of each mathematical concept such as the symbols, language, and structure. Students should identify the math language and symbols to determine what is being asked of them and use this information to determine which strategies to use to complete the task. Although it is not necessary for students to fill out a graphic organizer for each math problem they attempt, completing the following graphic organizer is helpful for annotating and reviewing math concepts and choosing appropriate strategies to complete tasks. The following is an example of how a student (with guidance) might fill out this graphic organizer:

	Symbols	Language	Operation or Action	Structure	Strategies/Tools
Part 1 Fractions vs. Decimals	. $\frac{\blacksquare}{\blacksquare}$	place value: tens/ tenths; fraction, numerator, denominator; < > =	compare	numerator over fraction bar/ denominator under; denominator = number of decimal places	use a number line to order decimals and fractions; use denominator to determine decimal places & vice versa
Part 2 Forms of Fractions	$\frac{\blacksquare}{\blacksquare}$	proper, improper, mixed number, remainder,	multiply and divide	numerator over fraction bar/ denominator under; whole number on the left, fraction on the right	fraction bar = divide; use division to change improper fraction to a mixed number/ multiplication to change mixed number to improper fraction
Part 3 Equivalent Fractions	× = ÷ $\frac{\blacksquare}{\blacksquare}$	higher/lower terms, raise, simplify, cross multiply, cross product	multiply and divide	equal numerator & denominator = 1; equal cross products = equivalent fractions	cross multiplication to find equivalent fractions; use greatest common factor to simplify/least common multiple to raise (most useful)
Part 4 Comparing Fractions	< > + =	greater than, less than, equal to	compare = mental subtraction	need common denominators to compare	multiply denominators or find least common multiple; cross multiply as short cut
Part 5 Add/ Subtract Fractions	+ − = $\frac{\blacksquare}{\blacksquare}$	add, and; reduce, how much faster; like, unlike	add, subtract	denominator stays the same, add/ subtract numerator across.	line up like fractions. move denominator into answer; calculate numerator and move to answer

2. Providing a real life context for a math concept is an important way to build conceptual understanding. Commonly divided shapes are often used to build conceptual understanding of fractions. *Math Sense 1: Focus on Operations* uses boxes with shaded areas to represent the fraction. Food is a great way to get students to understand the importance of fractions, especially if it is food they like. Due to a common fear of fractions, giving students opportunities to physically manipulate fractions in a variety of ways may help increase their conceptual understanding, build confidence in manipulating fractions, and as a by-product, alleviate this fear. Here is one example of how you may include this in a lesson using low cost items:

Set up five stations around the room, each with a food that can easily be broken into fractions.

Directions:

1. Start at one station and work at that station for 5 minutes before moving to the next station.
2. Follow the directions at each station.
3. Create each of the four fractions.
4. Draw each fraction you create in your notebook.

Fraction Stations:

1. ***Tortilla Slices:*** cut tortillas into the following fractions: $\frac{1}{2}$, $\frac{1}{3}$, $\frac{3}{4}$, $\frac{5}{8}$
2. ***Mandarin Oranges:*** divide into the following fractions: $\frac{1}{4}$, $\frac{2}{8}$, $\frac{1}{2}$, $\frac{4}{8}$
3. ***Water Measurement:*** measure out the following fractions: $\frac{1}{2}$ cup, $\frac{2}{3}$ cup, $\frac{1}{4}$ cup, $\frac{4}{8}$ cup
4. ***M&Ms Eating:*** start with 10 each time, eat the following fractions: $\frac{1}{10}$, $\frac{5}{10}$, $\frac{3}{10}$, $\frac{8}{10}$
5. ***Toothpick Cutting:*** cut into the following fractions: $\frac{1}{2}$, $\frac{4}{8}$, $\frac{1}{4}$, $\frac{5}{10}$

This example shows a number of equivalent fractions. After the exercise, bring students' attention to these equivalent fractions and explore the math behind them. Students should use their notebook illustrations to compare fractions. Consider the myriad of ways to extend concept building using these student notes.

3. *Math Sense 1: Focus on Operations* provides a Decimal, Fraction, and Percent Equivalencies chart (p. 205) that contains the most common decimals and their fraction and percent equivalencies. Students can use this resource to remind them how to represent these equivalencies in the various mathematical ways. In the *Bridging* Unit 2, Section 1 lesson (p. 21), students created posters to illustrate a variety of ways to represent common decimal amounts. For this lesson, students can return to these posters and add the equivalent fractions. They can also add more examples that are more typically represented by fractions, such as pizzas, oranges, filling up the gas tank, etc. Make sure that again students label each additional example with the unit and the whole it represents. For example: unit = tanks of gas; whole = 1 tank of gas.

4. The poster example above provides an excellent way to evaluate student understanding. In addition, students can explain to other students the choices they made for their poster and why. This can be accomplished through a jigsaw activity. Have pairs of students team up with one or two other pairs. Each pair takes a turn explaining their poster, describing their examples and the mathematical reasons for their choices. Stress to students the importance of using precise mathematical language in their explanations and supporting them with sound mathematical reasoning. A good way to set up this exchange is to provide a checklist of student expectations.

- ☑ Introduce your poster with the title (include fractions in addition to decimals).
- ☑ Take turns with your partner describing each of the examples.
- ☑ Use precise mathematical language, including saying your fraction correctly.
- ☑ Explain how your example mathematically represents the fraction (your mathematical reasoning).

➢ Bridging Vocabulary

Strategy 1: ***Identify the component parts and usage of new words to interpret their meanings.***

Strategy 2: ***Use context clues to interpret new words.***

Strategy 3: ***Utilize vocabulary-building resources.***

Strategy 4: ***Build a deeper knowledge of words through math application tasks and collaborative discussions.***

Strategy 5: ***Memorize words through repetitive study such as using flashcards (digital or print) and notes.***

1. First, present the shortest form of the word, referred to in this text as the "base word" in the case of academic words and some subject-specific terms. Follow the base form with other commonly used word forms (if available). Examine prefixes and suffixes and their impact on word meaning and part of speech.
2. Read the word as used in the context of the text and discuss possible meanings given context clues and word form.
3. Have students find (electronically or in print) the definition or translation of the base form and, if different, the form used in context and note these definitions in the space provided for future reference and study.
4. Gradually build a deeper knowledge of the word by having students use the word in a sentence frame, guided discussion, and an original sentence within a mathematical context.

Sentence Frame:	*The **fraction bar** separates the __________________ from the __________________.*
Guided Discussion:	*What is the relationship between the **numerator** and the **denominator**?*
Original Math Sentence:	____

Encourage students to use these words in math applications and collaborative discussions such as the task described in Bridging Problem Solving, Strategy 4.

5. The high volume of mathematical terminology requires repeat exposure to these words over time. Word walls, intentionally including the words in questions to students and when eliciting responses from them, and explicit reminders to use the vocabulary in verbal tasks provide built-in reinforcement. However, this is often not enough, so it is important that students learn ways to study words independently. Flashcards or websites that offer repetitive vocabulary practice are excellent ways for students to do this. Students may also use their notes, however, they will need to do repetitive activities, similar to flashcard practice, and not simply read and reread their notes.

➢ Bridging Math Application

Strategy 1: ***Prepare for math applications by identifying the problem type and the problem-solving strategies and tools.***

Strategy 2: ***Organize the problem using visual, symbolic, and written representations.***

Strategy 3: ***Overcome barriers to problem solving using math models, language and structural analysis, and resources.***

Strategy 4: ***Demonstrate and defend problem-solving application and mathematical reasoning through reverse problem solving, mental mathematics, visual representations, and peer discussions.***

1. Each part of this lesson lends itself to a variety of math application tasks that allow students to synthesize, apply, or extend their mathematical knowledge and skills. Whichever math application task you choose, be sure to orient students to the problem type and the problem-solving strategies and tools they may utilize. The following activity, synthesizes addition and subtraction into one real-life application.

Healthy Bridge Mix

In groups of 3 or 4, follow the recipe to make the Healthy Bridge Mix for your group. Each person must measure his or her own individual serving of each ingredient into a large bowl. Together, calculate how much of each ingredient was used by the whole group. Also calculate the total amount of all ingredients used. Finally, take out four ¼ cups of bridge mix and place in a separate cup for your teacher. Calculate how much bridge mix is left in the bowl for your group.

Recipe: (Individual Serving)

Materials
¼ cup measuring cup*
1 two-quart bowl
1 large plastic cup

Ingredients**
popped popcorn
sunflower seeds (without shells)
chocolate chips (or M&Ms)
raisins

Directions

Combine the following ingredients into the large bowl:

- ¾ cup popcorn
- ¼ cup sunflower seeds (without shells)
- ¼ cup chocolate chips (or M&Ms)
- ¼ cup raisins

* This can be a small paper cup. Measure ¼ cup and label in permanent marker.

**Using different colors of confetti is a low-cost way to complete the same activity…though somewhat less satisfying!

Before engaging in problem solving, have students analyze the directions to determine what is being asked of them and use this information to determine which strategies to use to complete the task. The following chart begins this process to provide a glimpse of how this may look.

Directions	Operation or Action *Signal Words*	Strategies and Tools
Each person must measure their own individual serving of each ingredient into a large bowl.	Measure	Use ¼ cup measuring cup. Fill to the top or to the line for each ¼ cup needed. Count the number of ¼-cup measures needed for a serving.
Together, calculate how much of each ingredient was used by the whole group.		

2. Encourage students to create their own graphic organizer that outlines the steps they need to take and gives them a place to record their answers. This may take a lot of guidance to lead them toward an effective plan. For students who struggle with organizing, provide a template but gradually release responsibility as they get more exposure to project planning.

Activity: Healthy Bridge Mix

Problem Steps

1. Each person measures an individual serving (according to the recipe) into the bowl.
2. Add the total amount of each ingredient used by the group.

	Add:		Total:
¾ cup popcorn	________	=	________
¼ cup sunflower seeds (without shells)	________	=	________
¼ cup chocolate chips (or M&Ms)	________	=	________
¼ cup raisins	________	=	________

3. Add the total amount of *all* ingredients used. Total: ________
4. Take out *four* ¼ cups and place in the plastic cup. Subtract: ________
5. Subtract this to find the amount left in the bowl. * New Amount: ________

* Let students know that the actual amount of Bridge Mix left in the bowl may not match their calculations exactly due to settling of the ingredients.

3. Students can use the tools and strategies they have learned leading up to this task, like the *Math Sense 1: Focus on Operations* lessons that feature visuals of fractions, to show the how and why of adding like fractions.
4. Once students have completed the task, have them meet with another group to compare their calculations. Students should choose a strategy to defend their calculations such as reverse problem solving, mental math, or using a visual representation.

➢ Assessment & Next Steps

Students should complete the practice activities included in each *Math Sense 1: Focus on Operations* lesson. Evaluate which learning goals were not met and remediate by using other resources, such as those identified in the Bridging Knowledge section. Upon successful completion, continue to the next section of the unit.

FRACTIONS: Section 2

Skills-Based Questions:

1. Why do you need common denominators to add and subtract fractions? How do you find them? What is the least common multiple (LCM) and why is it useful? *(Part 6)*
2. How do you add and subtract unlike fractions? *(Part 7)*
3. How do you add and subtract mixed numbers? Why might you need to regroup and how do you do that? *(Part 8)*
4. How do you multiply fractions? How do you cancel out common multiples and why is this useful? *(Part 9)*
5. How do you divide fractions? Why does changing the operation from division to multiplication work? *(Part 10)*
6. How do you multiply and divide mixed numbers? Why is it necessary to change mixed numbers to improper fractions before you multiply or divide? *(Part 11)*

Math Sense 1: Focus on Operations: Part 6, p. 80; Part 7, p. 82; Part 8, p. 84; Part 9, p. 88; Part 10, p. 90; Part 11, p. 92

	Learning Goals:	GED
Knowledge Goals:	1. Explain why a common denominator is necessary for adding and subtracting fractions. *(Part 6)*	Q.1.b Q.2.a
	2. Describe the least common multiple and how it is useful for adding and subtracting unlike fractions. *(Parts 6–7)*	
	3. Explain why regrouping is sometimes necessary when adding and subtracting mixed numbers. *(Part 8)*	
	4. Describe why multiplying factions results in a smaller number. *(Part 9)*	
	5. Explain the relationship between multiplication and division and how understanding that relationship helps you divide fractions. *(Part 10)*	
	6. Describe how to multiply and divide mixed numbers and why it is necessary to first change them to improper fractions. *(Part 11)*	
Problem-Solving Goals:	1. Find common denominators and least common multiples. *(Part 6)*	Q.1.b
	2. Add and subtract unlike fractions using the least common multiple. *(Part 7)*	Q.2.a
	3. Add and subtract mixed numbers using regrouping when necessary. *(Part 8)*	Q.2.e
	4. Multiply fractions canceling out common multiples when possible. *(Part 9)*	
	5. Divide fractions. *(Part 10)*	
	6. Multiply and divide mixed numbers by first changing them to improper fractions. *(Part 11)*	
Vocabulary Goals:	1. Define key mathematical terms.	
	2. Determine the meaning of unknown vocabulary using context clues, word forms, and parts of speech.	
	3. Apply new vocabulary to mathematical tasks and discussions.	
Math Application Goals:	1. Apply conceptual understanding of fractions to real-life situations. *(Parts 6–11)*	Q.1.b
	2. Defend math applications and reasoning to others. *(Parts 6–11)*	Q.2.a Q.2.e

Sample Instructional Support Strategies

➢ Bridging Knowledge

Strategy 1: ***Develop and connect background knowledge, skills, and conceptual understanding to new knowledge.***

Strategy 2: ***Use guiding questions to make connections beyond the lesson to broader math applications.***

Strategy 3: ***Use problem-solving strategies to develop, monitor, and synthesize conceptual understanding and fluency.*** *(See also Bridging Problem Solving)*

Strategy 4: ***Extend problem-solving skills and mathematical reasoning to broader math applications in life and work.*** *(See Bridging Math Application)*

1. Evaluate students' knowledge of the following mathematical skills. Utilize the chart below to develop student content knowledge as necessary.

Unit 3 **Fractions, Section 2**	**Part 6** Finding Common Denominators	**Part 7** Adding and Subtracting Unlike Fractions	**Part 8** Working with Whole and Mixed Numbers	**Part 9** Multiplying Fractions	**Part 10** Dividing Fractions	**Part 11** Multiplying and Dividing With Mixed Numbers
Core Skills in Mathematics	Unit 1, Lesson 5: Finding Common Factors and Multiples (p. 27)	Unit 1, Lesson 7: Performing Operations on Fractions and Mixed Numbers (p. 35)				
Scoreboost Mathematics: Fractions, Decimals, Percents, and Proportions	Decide Which Operation to Use with Fractions (p. 18); Solve Fraction Problems Using the Calculator (p. 22)					
Pre-HSE Workbook: Math 1	Adding and Subtracting Fractions, p. 16			Multiplying and Dividing Fractions (p. 18)		

2. The unit preview for each unit in the *Math Sense 1: Focus on Operations* provides a list of context examples in which to apply the target math concepts as well as questions that connect to students' prior knowledge and experience (for Unit 3, p. 67). These questions also provide a bridge to broader math applications in life and work.

3. Each unit also provides specialized lessons that focus on problem solving, using tools, and test taking techniques using the math skills taught in the lesson. Utilize these lessons to build student knowledge in these areas.

Unit 3, Section 2	Content	Page
Problem Solver	Multiples and Factors	78
	Does the Answer Make Sense?	94
Tools	Calculators and Fractions	96
Test Taker	Know When to Use Decimals	98

➢ Bridging Problem Solving

Strategy 1: ***Preview the problem to determine problem-solving strategies and tools and predict general solutions.***

Strategy 2: ***Develop conceptual understanding of mathematical problems using visual representations, think-alouds, and collaboration.***

Strategy 3: ***Overcome barriers to problem solving using math models, language and structural analysis, and resources.***

Strategy 4: ***Demonstrate and defend problem solving and mathematical reasoning through reverse problem solving, mental mathematics, visual representations, and peer discussions.***

1. For Parts 6–11 of Unit 3, orient students to features of each mathematical concept such as the symbols, language, and structure. Students should identify the math language and symbols to determine what is being asked of them and use this information to determine which strategies to use to complete the task. Although it is not necessary for students to fill out a graphic organizer for each math problem they attempt, completing the following graphic organizer is helpful for annotating and reviewing math concepts and choosing appropriate strategies to complete tasks. The following is an example of how a student (with guidance) might fill out this graphic organizer:

	Symbols	Language	Operation or Action	Structure	Strategies/Tools
Part 6 Common Denominators	■/■ + − × ÷ =	common denominator, least common multiple	multiply	denominators for each fraction must be equal to perform addition or subtraction	multiply denominators or use LCM to get common denominator
Part 7 Add/ Subtract Unlike Fractions	■/■ + − =	numerator, common denominator, unlike fractions, equivalent fractions, simplify	add and subtract	fractions must have equal denominators; line up across	use LCM to get least common denominator; use equivalent fractions to set up equation
Part 8 Work with Whole and Mixed Numbers	■/■ + − =	whole number, mixed number, common denominator, equivalent fractions	add and subtract; multiply	whole number on left, fraction on right; line up and down for equation	add/subtract fractions first; add whole numbers; add regrouped fraction to whole number when necessary
Part 9 Multiply Fractions	■/■ × =	cancel, common factors, simplify	multiply	set up fractions across from each other	cancel common factors; multiply across; simplify answer by dividing by common factor if necessary
Part 10 Divide Fractions	■/■ × ÷ =	reciprocal, invert, cancel, common factors, simplify	divide, invert, multiply	invert the divisor, change division to multiplication	multiply reciprocal of divisor (division = opposite of multiplication; reciprocal = opposite of divisor)
Part 11 Multiply/ Divide Mixed Numbers	■/■ × ÷ =	mixed numbers, improper fractions, cancel, reciprocal, invert, simplify	multiply and divide	mixed numbers change to improper fractions; fractions line up across from each other	multiply mixed numbers to get improper fractions; multiply numerators and denominators; divide to change answer to a mixed number

2. Providing a real-life context is an important way to build conceptual understanding. Commonly divided shapes are often used to build conceptual understanding of fractions. *Math Sense 1: Focus on Operations* uses shaded boxes or pie wedges to represent fractions. Food is a great way to get students to understand the importance of fractions, especially if it is food they like. Due to a common fear of fractions, giving students opportunities to physically manipulate fractions in a variety of ways may help increase their conceptual understanding, build confidence in manipulating fractions, and as a by-product, alleviate this fear. The previous lesson provided an activity to conceptualize fractions. That activity can be adjusted to practice using the four operations with unlike fractions.

Set up five stations around the room each with a food that can easily be broken into fractions.

Directions:

1. Start at one station and work at that station for 10 minutes before moving to the next station.
2. Follow the directions at each station.
3. Create each of the four fractions.
4. Answer the question(s) by using the appropriate mathematical operations.

Fraction Stations:

1. ***Tortilla Slices:*** cut tortillas into the following fractions: $\frac{1}{2}$, $\frac{1}{3}$, $\frac{3}{4}$, $\frac{5}{8}$
   How many total tortillas did you cut off? ____________
2. ***Mandarin Oranges:*** break into the following fractions: $\frac{1}{4}$, $\frac{2}{8}$, $\frac{1}{2}$, $\frac{4}{8}$
   How many total oranges did you break off? ____________
   If someone eats the $\frac{2}{8}$ section how many of the oranges you broke off are left? ____________
3. ***Water Measurement:*** measure out the following fractions: $\frac{1}{2}$ cup, $\frac{2}{3}$ cup, $\frac{4}{4}$ cup, $\frac{4}{8}$
   How much total water did you measure? ____________
   If you want to give 3 people each $\frac{1}{3}$ of the total, how much will each person get? ____________
4. ***M&Ms Eating:*** start with 10 each time, eat the following fractions: $\frac{1}{10}$, $\frac{5}{10}$, $\frac{3}{10}$, $\frac{8}{10}$
   How many total M&Ms did your group eat? ____________
   If you have eaten $\frac{1}{4}$ of the M&Ms from the bag, how many more times can you eat the same amount? ____________

5. During the above activity, students can use the lessons in *Math Sense 1: Focus on Operations* Unit 3: Fractions to remind them of the steps necessary to solve each problem. Or students can create a cheat sheet of the steps to use prior to the activity to guide their work.

6. Once students complete their calculations for each station, they should work together to check their answers. Have students estimate each of the fractions they are working with to the nearest whole number for larger fractions or half for smaller fractions and complete the problem using these estimates. Then, they can compare the answer they calculated using the actual fractions to see if their answer makes sense. Careful that students use the actual answer they checked for each subsequent step in the problem. Otherwise, their estimations may get further and further from the actual answer. For example:

Actual: $\frac{1}{4} + \frac{2}{8} + \frac{1}{2} + \frac{4}{8} = x$	Estimated: $0 + 0 + \frac{1}{2} + \frac{1}{2} = x$
Actual: $1\frac{1}{2}$	Estimated: 1
Actual: $1\frac{1}{2} - \frac{2}{8} = x$	Estimated: $1\frac{1}{2} - 0 = x$
Actual: $1\frac{1}{4}$	Estimated: $1\frac{1}{2}$

➢ Bridging Vocabulary

Strategy 1: Identify the component parts and usage of new words to interpret their meanings.

Strategy 2: Use context clues to interpret new words.

Strategy 3: Utilize vocabulary-building resources.

Strategy 4: Build a deeper knowledge of words through math application tasks and collaborative discussions.

Strategy 5: Memorize words through repetitive study such as using flashcards (digital or print) and notes.

1. First, present the shortest form of the word, referred to in this text as the "base word" in the case of academic words and some subject-specific terms. Follow the base form with other commonly used word forms (if available). Examine prefixes and suffixes and their impact on word meaning and part of speech.
2. Read the word as used in the context of the text and discuss possible meanings given context clues and word form.
3. Have students find (electronically or in print) the definition or translation of the base form and, if different, the form used in context and note these definitions in the space provided for future reference and study.
4. Gradually build a deeper knowledge of the word by having students use the word in a sentence frame, guided discussion, and an original sentence within a mathematical context.

Sentence Frame:	*The **reciprocal fraction** is the ____________________ of the ____________________.*
Guided Discussion:	*What is the relationship between the **reciprocal fraction** and the original fraction? When and why do we use the reciprocal fraction?*
Original Math Sentence:	____

Encourage students to use these words in math applications and collaborative discussions such as the task described in Bridging Problem Solving Strategy 4.

5. The high volume of mathematical terminology requires repeat exposure to these words over time. Word walls, intentionally including the words in questions to students and when eliciting responses from them, and explicit reminders to use the vocabulary in verbal tasks provide built-in reinforcement. However, this is often not enough so it is important that students learn ways to study words independently. Flashcards or websites that offer repetitive vocabulary practice are excellent ways for students to do this. Students may also use their notes, however, they will need to do repetitive activities, similar to flashcard practice, and not simply read and reread their notes.

➢ Bridging Math Application

Strategy 1: ***Prepare for math applications by identifying the problem type and the problem-solving strategies and tools.***

Strategy 2: ***Organize the problem using visual, symbolic, and written representations.***

Strategy 3: ***Overcome barriers to problem solving using math models, language and structural analysis, and resources.***

Strategy 4: ***Demonstrate and defend problem-solving application and mathematical reasoning through reverse problem solving, mental mathematics, visual representations, and peer discussions.***

1. Each part of this lesson lends itself to a variety of math application tasks that allow students to synthesize, apply, or extend their mathematical knowledge and skills. Whichever math application task you choose, be sure to orient students to the problem type and the problem-solving strategies and tools they may utilize. The following activity synthesizes the four operations with fractions into one real-life application.

Create a homemade window cleaning solution by combining the ingredients listed below. Then create a blog post that describes how to make this solution, the quantity of liquid it contains, and how to change the recipe to make two different quantities, a smaller quantity and a larger quantity.

Window Cleaner Recipe:

Combine: ½ cup water; ¼ cup club soda; ⅛ cup lemon juice

Before engaging in problem solving, have students analyze the directions to determine what is being asked of them and use this information to determine which strategies to use to complete the task. The following chart begins this process to provide a glimpse of how this may look.

Directions	Operation or Action *Signal Words*	Strategies and Tools
Create a homemade window cleaning solution by combining the ingredients listed below.	Measure	Use measuring cups to measure ingredients into a container.
Then create a blog post that describes how to make this solution …		

2. Encourage students to create their own graphic organizer that outlines the steps they need to take and gives them a place to record their answers. This may take a lot of guidance to lead them toward an effective plan. For students who struggle with organizing, provide them with a template but gradually release responsibility as they get more exposure to project planning.

Blog Post Task

1. Make Recipe		2. Greater Quantity	3. Smaller Quantity
Steps to write: 1. ________ 2. ________ 3. ________	Total Quantity: Measurements: ________ with LCM: ________ Total: ________	Multiply each measurement by 2 to make a greater quantity 1. Water: ______ = ______ 2. Club soda: ______ = ______ 3. Lemon Juice: ______ = ______	Multiply each measurement by ½ to make a smaller quantity. 1. Water: ______ = ______ 2. Club soda: ______ = ______ 3. Lemon Juice: ______ = ______

3. During the above activity, students can use the lessons in *Math Sense 1: Focus on Operations* Unit 3, Fractions to remind them of the steps necessary to solve each problem. Or students can create a cheat sheet of the steps to use prior to the activity to guide their work.
4. Once students have written their blog post, have them post their blogs on your class web page. Have students comment on at least one other blog about the content of the blog as well as the accuracy of the calculations.

➤ Assessment & Next Steps

Students should complete the practice activities included in each *Math Sense 1: Focus on Operations* lesson. Evaluate which learning goals were not met and remediate by using other resources, such as those identified in the Bridging Knowledge section. Upon successful completion, continue to the next unit.

Unit 4

RATIO, PROPORTION, AND PERCENT: Section 1

Skills-Based Questions:

1. How are ratios and fractions alike? What are three ways to express a ratio? *(Part 1)*
2. How do you set up a ratio? *(Part 2)*
3. How do you find the information you need to calculate problems using ratios? *(Part 3)*
4. What is the difference between a ratio and a proportion? How do you solve for a missing number in a proportion? *(Parts 4–5)*

Math Sense 1: Focus on Operations: Part 1, p. 104; Part 2, p. 106; Problem Solver, p. 108; Part 3, p. 110; Parts 4–5, p. 112–115

	Learning Goals:	**GED**
Knowledge Goals:	1. Describe what a ratio is and its relationship to a fraction. *(Part 1)*	Q.3.a
	2. Explain how to set up a ratio in three different ways. *(Part 2)*	Q.3.b
	3. Describe how to determine a unit rate given a ratio. *(Problem Solver, p. 108)*	Q.3.c
	4. Describe the given information in a problem and explain how to use it to find missing information needed to solve a problem using ratios. *(Part 3)*	
	5. Describe the relationship between the two ratios in a proportion and how that relationship helps solve for a missing number in a proportion. *(Parts 4–5)*	
Problem-Solving Goals:	1. Use ratios to find fractions when comparing a part to a whole. *(Part 1)*	Q.3.a
	2. Set up ratios appropriately in three different ways. *(Part 2)*	Q.3.b
	3. Calculate unit rates from given ratios. *(Problem Solver, p. 108)*	Q.3.c
	4. Solve problems using ratios, including problems that require finding information and changing amounts. *(Part 3)*	
	5. Set up proportions and solve for missing numbers. *(Parts 4–5)*	
Vocabulary Goals:	1. Define key mathematical terms.	
	2. Determine the meaning of unknown vocabulary using context clues, word forms, and parts of speech.	
	3. Apply new vocabulary to mathematical tasks and discussions.	
Math Application Goals:	1. Apply conceptual understanding of ratios and proportions to real life situations. *(Parts 1–5)*	Q.3.a
	2. Defend math applications and reasoning to others. *(Parts 1–5)*	Q.3.b
		Q.3.c

Sample Instructional Support Strategies

➢ Bridging Knowledge

Strategy 1: ***Develop and connect background knowledge, skills, and conceptual understanding to new knowledge.***

Strategy 2: ***Use guiding questions to make connections beyond the lesson to broader math applications.***

Strategy 3: ***Use problem-solving strategies to develop, monitor, and synthesize conceptual understanding and fluency.*** *(See also Bridging Problem Solving)*

Strategy 4: ***Extend problem-solving skills and mathematical reasoning to broader math applications in life and work.*** *(See Bridging Math Application)*

1. Evaluate students' knowledge of the following mathematical skills. Utilize the chart below to develop student content knowledge as necessary.

Unit 4 **Ratio, Proportion, and Percent, Section 1**	**Part 1** Relating Fractions and Ratios	**Part 2** Writing Ratios	**Problem Solver** Unit Rates	**Part 3** Ratios in Word Problems	**Part 4** Writing Proportions	**Part 5** Solving Problems with Proportions
Core Skills in Mathematics	Unit 2, Lesson 1: Using Ratios and Proportions to Solve Problems (p. 44)					
Scoreboost Mathematics: Fractions, Decimals, Percents, and Proportions	Solve Ratio and Proportion Problems (p. 28)					
Pre-HSE Workbook: Math 1	Solving Ratio and Proportion Problems (p. 24)					

2. The unit preview for each unit in the *Math Sense 1: Focus on Operations* provides a list of context examples in which to apply the target math concepts as well as questions that connect to students' prior knowledge and experience (for Unit 4, p. 103). These questions also provide a bridge to broader math applications in life and work.

3. Each unit also provides specialized lessons that focus on problem solving, using tools, and test taking techniques using the math skills taught in the lesson. Utilize these lessons to build student knowledge in these areas.

Unit 4, Section 1	**Content (GED Target)**	**Page**
Problem Solver	Unit Rates (Q.3.a) Scale Drawings (Q.3.b)	108 118
Tools	Using a Calculator with Proportion	116
Test Taker	NA	

➢ Bridging Problem Solving

Strategy 1: ***Preview the problem to determine problem-solving strategies and tools and predict general solutions.***

Strategy 2: ***Develop conceptual understanding of mathematical problems using visual representations, think-alouds, and collaboration.***

Strategy 3: ***Overcome barriers to problem solving using math models, language and structural analysis, and resources.***

Strategy 4: ***Demonstrate and defend problem solving and mathematical reasoning through reverse problem solving, mental mathematics, visual representations, and peer discussions.***

1. For Parts 1–5 of Unit 4, orient students to features of each mathematical concept such as the symbols, language, and structure. Students should identify the math language and symbols to determine what is being asked of them and use this information to determine which strategies to use to complete the task. Although it is not necessary for students to fill out a graphic organizer for each math problem they attempt, completing the following graphic organizer is helpful for annotating and reviewing math concepts and choosing appropriate strategies to complete tasks. The following is an example of how a student (with guidance) might fill out this graphic organizer:

	Symbols	Language	Operation or Action	Structure	Strategies/Tools
Part 1 Ratios and Fractions	$\frac{\blacksquare}{\blacksquare}$:	ratio, fraction, part, whole, compare, 'to,' simplify, lowest terms	compare	fraction (numerator over the bar/denominator below), ':' or 'to' separating units	first unit in problem = first unit in ratio, label units; a fraction is when comparing part to whole
Part 2 Writing Ratios	$\frac{\blacksquare}{\blacksquare}$:	ratio, ratio of, 'to,' total, units (width, length, height, minutes, etc.)	set up	fraction (numerator over the bar/denominator below), ':' or 'to' separating units	first unit in problem = first unit in ratio, label units
Problem Solver Unit Rates	÷ = $\frac{\blacksquare}{\blacksquare}$	unit rate, ratio, compare, 'per,' for each	divide	unit to divide (dividend) = numerator / unit to divide by (divisor) = denominator	divide numerator and denominator by the denominator (denominator same on top and bottom, so same as dividing numerator by denominator); label units
Part 3 Solving Ratio Problems	+ − ÷ × = $\frac{\blacksquare}{\blacksquare}$	compare, ratio, quantity, missing information (add or subtract), how many added	add, subtract (missing info)	equal numerator and denominator = 1; equal cross products = equivalent fractions	find total first; identify two or more variables to be compared; only 1? subtract from total to find the other variable; use knowledge of fractions to find new ratio when quantity of variables change
Parts 4–5 Writing Proportions	÷ × = $\frac{\blacksquare}{\blacksquare}$	proportion, ratio, equivalent fractions, cross products, equations, variable,	multiply and divide	set up equivalent fractions with = between (equation); use a letter for missing variable	two equal ratios = proportion; cross multiply to find equivalent fractions (cross products); use a table to set up variables; set up equation and use cross products to solve for missing variable; label units

2. Utilizing student information and experiences to explore math concepts creates a concrete way to see a math concept in action as well as promotes collaboration at the same time. Create a survey with five or more questions that can be used to compare data through ratios. Have students gather the answers to the questions from their classmates (or from outside sources). Once they have collected the data, they can explore the relationships between the results through guided questions that require ratios.

Example Survey Questions:

Ask your classmates the following questions. Make sure to check male or female when you record an answer.

M	F	Do you like to play sports?	Do you have children?	Do you plan to go to college?	What is your favorite color?	Do you study at home 10 hours or more per week?

Write a ratio for each question:

1. What is the ratio of men to women who like to play sports? ____________
2. What is the ratio of students who like to play sports to total students? ____________
3. What is the ratio of men to women who have children? ____________
4. What is the ratio of students who have children to total students? ____________
5. What is the ratio of women to men who plan to go to college? ____________
6. What is the ratio of students who plan to go to college to total students? ____________
7. What is the ratio of men to women who like the color green the most? ____________
8. What is the ratio of women to men who study 10 hours or more per week? ____________
9. What is the ratio of students who study 10 hours or more per week to total students? ____________
10. What is the ratio of students who study less than 10 hours per week to students who study 10 hours or more? ____________

3. One very important aspect of ratios and proportions is to keep what is being compared in the correct order. Analyzing the structure of a word problem or question and then following that structure in one's answer can help students focus on the precision required for accurate mathematical computations. For example, for the above ratio questions, students must be able to analyze the question to identify the variables they need to consider, the values/answers for those variables, and often the total number. A simple checklist may help them identify these within the problem.

1. What variables do I need to look at?
2. What is the value (answer) for each of those variables?
3. Do I need to know the total to find the information I need? If so, do I need to subtract from the total?

4. Have students work in groups to analyze the data collected from the surveys. They should use this as an opportunity to explain their answers and probe the answers of others. This holds them accountable for their mathematical reasoning and helps them maintain a discerning ear to the explanations of others. Using collaborative language is helpful when participating in this type of group work.

Polite disagreement: *I see what you're saying but I think ______________. Another way to think about this is ______________.*

Encourage participation: *What do you think? I'd like to hear what you have to say about ______________.*

Probe others' ideas: *Could you explain that further? What did you mean by ______________?*

Paraphrase: *So, do we agree that ______________? I think you're saying that ______________.*

Present examples: *An example of this is ______________. ______________ is an example of ______________.*

➢ Bridging Vocabulary

Strategy 1: Identify the component parts and usage of new words to interpret their meanings.

Strategy 2: Use context clues to interpret new words.

Strategy 3: Utilize vocabulary-building resources.

Strategy 4: Build a deeper knowledge of words through math application tasks and collaborative discussions.

Strategy 5: Memorize words through repetitive study such as using flashcards (digital or print) and notes.

1. First, present the shortest form of the word, referred to in this text as the "base word" in the case of academic words and some subject-specific terms. Follow the base form with other commonly used word forms (if available). Examine prefixes and suffixes and their impact on word meaning and part of speech.
2. Read the word as used in the context of the text and discuss possible meanings given context clues and word form.
3. Have students find (electronically or in print) the definition or translation of the base form and, if different, the form used in context and note these definitions in the space provided for future reference and study.
4. Gradually build a deeper knowledge of the word by having students use the word in a sentence frame, guided discussion, and an original sentence within a mathematical context.

Sentence Frame:	*A **proportion** shows ______________.*
Guided Discussion:	*How do you solve for a missing number in a **proportion**?*
Original Math Sentence:	______________________________

Then encourage students to use these words in math applications and collaborative discussions such as the task described in Bridging Problem Solving, Strategy 4.

5. The high volume of mathematical terminology requires repeat exposure to the words over time. Word walls, intentionally including the words in questions to students and when eliciting responses from them, and explicit reminders to use the vocabulary in verbal tasks provide built-in reinforcement. However, this is often not enough so it is important that students learn ways to study words independently. Flashcards or websites that offer repetitive vocabulary practice are excellent ways for students to do this. Students may also use their notes, however, they will need to do repetitive activities, similar to flashcard practice, and not simply read and reread their notes.

➢ Bridging Math Application

Strategy 1: ***Prepare for math applications by identifying the problem type and the problem-solving strategies and tools.***

Strategy 2: ***Organize the problem using visual, symbolic, and written representations.***

Strategy 3: ***Overcome barriers to problem solving using math models, language and structural analysis, and resources.***

Strategy 4: ***Demonstrate and defend problem-solving application and mathematical reasoning through reverse problem solving, mental mathematics, visual representations, and peer discussions.***

1. Each part of this lesson lends itself to a variety of math application tasks that allow students to synthesize, apply, or extend their mathematical knowledge and skills. Whichever math application task you choose, be sure to orient students to the problem type and the problem-solving strategies and tools they may utilize. The following activity, synthesizes the use of ratios and proportions into a real-life application.

Team Bracelets

In groups of three or four, make team bracelets for your group. First follow the directions below to make a prototype, or test bracelet. Find the ratio of the length of each string to the length of the finished bracelet. Then, use the ratio to set up a proportion that will allow you to calculate the length of string you need for each team member's finished bracelet. You will need to measure your teammates' wrists to know how long each finished bracelet should be. (Note: add 4 inches to this length for tying the finished bracelet.) Design an equation to find the missing variable in each proportion. Finally, follow the directions using the lengths you calculated to make a bracelet for each member of your team.

Materials:

A variety of colored strings of the same ply

A flexible measuring tape

Scissors

Prototype Directions:

1. Cut 5 strings, each 16 inches in length (Your group can choose which colors to use.)
2. Fold strings in half and tie all strings together at the fold. (Make a loop and pull the folded ends through the loop. Pull tight.)
3. Separate strings into 5 strands using same-colored pairs. Each color will move as one strand.
4. Follow the teacher's (or video's) directions on how to make a 5-strand braid.
5. Continue the 5-strand braid until you have just enough string left to tie the loose ends together in a double knot.

Before engaging in problem solving, have students analyze the directions to determine what is being asked of them and use this information to determine which strategies to use to complete the task. The following chart begins this process to provide a glimpse of how this may look.

Directions	**Operation or Action** *Signal Words*	**Strategies and Tools**
First follow the directions below to make a prototype, or test bracelet.	cut and measure strings fold and tie strings separate and braid strings tie at end	Use a measuring tape to measure strings; follow directions step-by-step
Find the ratio of the length of each string to the length of the finished bracelet.		

2. Students will need to plan and organize how to complete the previous task. Encourage students to create their own graphic organizer that outlines the steps they will need to take and gives them a place to record their answers. This may take a lot of guidance to lead them toward an effective plan. For students who struggle with organizing, provide them with a template but gradually release responsibility as they get more exposure to project planning.

Activity: Team Bracelets

Materials:

- A variety of colored strings of the same ply
- A flexible measuring tape
- Scissors

Prototype Directions:

1. Cut 5 strings, each 16 inches in length.
2. Fold strings in half and tie all strings together at the fold.
3. Separate strings into 5 strands using same-colored pairs. Each color will move as one strand.
4. Follow directions on how to make a 5-strand braid.
5. Continue the braid until you have just enough string to tie in a double knot.

Set up the Ratio:

1. Write down the length of the string. ____________
2. Measure and write down the length of the finished bracelet. ____________
3. Create a proportion that shows the length of the string to the length of the finished bracelet.
   ____________ (length of string) : ____________ (length of bracelet).
4. Estimate to check answer: Lay the bracelet next to the length of the string. Does it look like same ratio?

Team Bracelet Directions:

Use the ratio from the prototype to set up a proportion that will allow you to calculate the length of string you need for each teammates' finished bracelet.

	Length of string	**Length of bracelet**
Prototype		
Bracelet for *name*		
Bracelet for *name*		

Measure your teammates' wrists so that you know how long each finished bracelet should be. (Add 4 inches to this length.) Write the total length in the table under the correct variable.

Make an equation to find the missing variable in each proportion.

Finally, follow directions using the calculated variables to make a bracelet for each member of your team.

3. Students can use the tools and strategies they have learned leading up to this task, like the *Math Sense 1: Focus on Operations* lessons that feature step-by-step directions on how to set up ratios and proportions and how to solve for missing variables in a proportion.

4. Once students have completed the task, have them meet with another group to show their bracelets and compare their calculations. Students can defend their calculations by showing well-fitted bracelets or explain where their calculations went wrong, given ill-fitting bracelets.

➤ Assessment & Next Steps

Students should complete the practice activities included in each *Math Sense 1: Focus on Operations* lesson. Evaluate which learning goals were not met and remediate by using other resources, such as those identified in the Bridging Knowledge section. Upon successful completion, continue to the next section of this unit.

RATIO, PROPORTION, AND PERCENT: Section 2

Skills-Based Questions:

1. What do percents represent? *(Part 6)*
2. How are decimals, fractions, and percents alike? How do you express the same value as a decimal, fraction, and percent? *(Part 7)*
3. How do you set up a percent equation? What variables does a percent equation allow you to solve for? *(Part 8)*
4. How do you solve a percent equation for any one missing variable? *(Part 9)*
5. How do you solve two-step percent problems? *(Part 10)*
6. How do you determine a percent increase or decrease? *(Part 11)*

Math Sense 1: Focus on Operations: Part 6, p. 122; Part 7, p. 124; Part 8, p. 126; Parts 9, p. 128; Part 10, p. 132; Part 11, p.138

	Learning Goals:	**GED**
Knowledge Goals:	1. Describe the relationship percents represent. *(Part 6)*	Q.3.d
	2. Describe the relationship between decimals, fractions, and percents and how one value can be expressed as each of these. *(Part 7)*	
	3. Explain how to set up a percent equation with the three variables it includes. *(Part 8)*	
	4. Explain how to solve a percent equation with one missing variable. *(Part 9)*	
	5. Describe how to solve a two-step percent problem. *(Part 10)*	
	6. Explain how you solve for a percent increase or decrease. *(Part 11)*	
Problem-Solving Goals:	1. Determine a percent value from a graphic or written problem. *(Part 6)*	Q.3.d
	2. Express a given value as a decimal, fraction, and a percent. *(Part 7)*	
	3. Set up percent equations by putting the whole, part, and percent in correct positions. *(Part 8)*	
	4. Solve percent equations for the missing variable (the whole, part, or percent). *(Part 9)*	
	5. Solve two-step percent problems. *(Part 10)*	
	6. Calculate a percent increase or decrease. *(Part 11)*	
Vocabulary Goals:	1. Define key mathematical terms.	
	2. Determine the meaning of unknown vocabulary using context clues, word forms, and parts of speech.	
	3. Apply new vocabulary to mathematical tasks and discussions.	
Math Application Goals:	1. Apply conceptual understanding of percents to real life situations. *(Parts 6–11)*	Q.3.d
	2. Defend math applications and reasoning to others. *(Parts 6–11)*	

Sample Instructional Support Strategies

➢ Bridging Knowledge

Strategy 1: ***Develop and connect background knowledge, skills, and conceptual understanding to new knowledge.***

Strategy 2: ***Use guiding questions to make connections beyond the lesson to broader math applications.***

Strategy 3: ***Use problem-solving strategies to develop, monitor, and synthesize conceptual understanding and fluency.*** *(See also Bridging Problem Solving)*

Strategy 4: ***Extend problem-solving skills and mathematical reasoning to broader math applications in life and work.*** *(See Bridging Math Application)*

1. Evaluate students' knowledge of the following mathematical skills. Utilize the chart below to develop student content knowledge as necessary.

Unit 4 **Ratio, Proportion, and Percent, Section 2**	**Part 6** Understanding Percents	**Part 7** Decimals, Fractions, and Percents	**Part 8** The Percent Equation	**Part 9** Solving Percent Equations	**Part 10** Two-Step Percent Problems	**Part 11** Percent of Increase/ Decrease
Core Skills in Mathematics	Unit 2, Lesson 2: Solving Percent Problems (p. 48)					
Scoreboost Mathematics: Fractions, Decimals, Percents, and Proportions	Use the Percent Circle (p. 30); Use the Calculator to Solve Percent Problems (p. 32)					
Pre-HSE Workbook: Math 1	Solving Percent Problems (p. 26)					

2. The unit preview for each unit in *Math Sense 1: Focus on Operations* provides a list of context examples in which to apply the target math concepts as well as questions that connect to students' prior knowledge and experience (for Unit 4, p. 103). These questions also provide a bridge to broader math applications in life and work.

3. Each unit also provides specialized lessons that focus on problem solving, using tools, and test taking techniques using the math skills taught in the lesson. Utilize these lessons to build student knowledge in these areas.

Unit 4, Section 2	Content	Page
Problem Solver	Common Percent Applications Use Proportion to Solve Percent Problems	130 136
Tools	Using a Calculator with Percents	134
Test Taker	Use Fractions to Solve Percent Problems	140

➤ Bridging Problem Solving

Strategy 1: ***Preview the problem to determine problem-solving strategies and tools and predict general solutions.***

Strategy 2: ***Develop conceptual understanding of mathematical problems using visual representations, think-alouds, and collaboration.***

Strategy 3: ***Overcome barriers to problem solving using math models, language and structural analysis, and resources.***

Strategy 4: ***Demonstrate and defend problem solving and mathematical reasoning through reverse problem solving, mental mathematics, visual representations, and peer discussions.***

1. For Parts 6–11 of Unit 4, orient students to features of each mathematical concept such as the symbols, language, and structure. Students should identify the math language and symbols to determine what is being asked of them and use this information to determine which strategies to use to complete the task. Although it is not necessary for students to fill out a graphic organizer for each math problem they attempt, completing the following graphic organizer is helpful for annotating and reviewing math concepts and choosing appropriate strategies to complete tasks. The following is an example of how a student (with guidance) might fill out this graphic organizer:

	Symbols	Language	Operation or Action	Structure	Strategies/Tools
Part 6 Understand Percents	%	percent, whole, part	subtract, add	100% = the whole less than 100% = part; 100% – part = complement part	identify whole (100%); identify part; subtract part from 100% to find complement part
Part 7 Decimals, Fractions, and Percents	. % ■/■	decimal point, fraction, numerator, denominator, percent	express as, move decimal point, write as	percent to decimal is 2 places left; fraction: part over fraction bar / whole (100%) under	change percent to decimal by moving decimal point 2 places left; opposite for decimal to percent; fill in places with zeros if necessary
Part 8 Percent Equations	× = % *p n w*	percent statement, equation, percent, whole, part, unknown, variable	change, multiply	percent of whole is part ($n\% \times w = p$)	change percent statement to percent equation; put variable in place of unknown
Part 9 Solving Percent Equations	÷ × = % *p n w*	percent equation, variable, unknown, part, whole	find, solve, multiply, divide	percent of whole is part ($n\% \times w = p$) $p = \% \times w$ $w = p/\%$ $\% = p/w$	change percent statement to percent equation; put variable in place of unknown; solve for variable by getting variable alone on one side of equation
Part 10 Two-Step Percent Problems	÷ × = % *p n w* – +	two-step, percent equation, variable	multiply and divide	percent of whole is part ($n\% \times w = p$)	solve the percent equation; use the answer in the next step of the problem; decide which operation to use or if another percent equation is needed
Part 11 Percent of Increase and Decrease	÷ × = % *p n w* – +	increase, decrease, part, whole, change, original	subtract, multiply, divide	the part = the change in value; the whole is the original value	find the change in value (increase or decrease) by subtracting; use that change as the part in the percent equation

2. Providing a real-life context for a math concept is an important way to build conceptual understanding. Divided shapes, such as squares and circles, are often used to build conceptual understanding of percents. *Math Sense 1: Focus on Operations* uses a square divided into 100 boxes with shaded boxes representing the percent and the unshaded boxes representing the complement to that percent (p. 122). *Scoreboost Mathematics: Fractions, Decimals, Percents, and Proportions* illustrates the relationship of the three parts of the percent formula using a percent circle (p. 30). Here is a fun (and delicious) way to build conceptual knowledge of percents and the use of a pie chart to represent percents.

Candy Colors

Materials:

- 1 large bag of Skittles or M&Ms* (enough for each student to have 100 pieces)
- 1 paper with a large circle in the middle (large enough to fit 100 pieces of candy)

Directions:

1. Pass out 100 pieces of candy of assorted colors to each student.
2. Have students separate the candy by color.
3. Students should count how many pieces of candy they have of each color and note it on a paper.
4. Using the circle on the paper, have students create pie wedges using each color within the circle.
5. On the outside of the circle, have students label the percent of each color pie wedge.
6. Next, have students turn to a partner and describe their pie chart and the percents of each colored candy.
7. Students can write the percents as decimals and fractions and share these expressions with their partner.
8. At the end of the activity, students may choose to eat their candy.

*Substitute colored beads for candy as a low-cost option that you can use again and again for this activity.

3. In step 6 of the above activity, students turn to a partner and describe their pie chart. This step is where students are responsible for explaining how they arrived at each percent. Students should focus on describing the important aspects of percents such as the part and the whole and how the whole equals 100%. In the second part of step 7, students turn to their partner and describe how they changed the percent to a decimal and a fraction. Again, the focus should be on using precise mathematical language to explain these actions.

➢ Bridging Vocabulary

Strategy 1: Identify the component parts and usage of new words to interpret their meanings.

Strategy 2: Use context clues to interpret new words.

Strategy 3: Utilize vocabulary-building resources.

Strategy 4: Build a deeper knowledge of words through math application tasks and collaborative discussions.

Strategy 5: Memorize words through repetitive study such as using flashcards (digital or print) and notes.

1. First, present the shortest form of the word, referred to in this text as the "base word" in the case of academic words and some subject-specific terms. Follow the base form with other commonly used word forms (if available). Examine prefixes and suffixes and their impact on word meaning and part of speech.
2. Read the word as used in the context of the text and discuss possible meanings given context clues and word form.
3. Have students find (electronically or in print) the definition or translation of the base form and, if different, the form used in context and note these definitions in the space provided for future reference and study.

4. Gradually build a deeper knowledge of the word by having students use the word in a sentence frame, guided discussion, and an original sentence within a mathematical context.

Sentence Frame:	*A **percent** is similar to a fraction because ________ but for a percent, the whole is always ________.*
Guided Discussion:	*In what situations do you use **percents**? Why are they useful in these situations?*
Original Math Sentence:	________________________________

Encourage students to use these words in math applications and collaborative discussions such as the task described in Bridging Problem Solving Strategy 4.

5. The high volume of mathematical terminology requires repeat exposure to the words over time. Word walls, intentionally including the words in questions to students and when eliciting responses from them, and explicit reminders to use the vocabulary in verbal tasks provide built-in reinforcement. However, this is often not enough so it is important that students learn ways to study words independently. Flashcards or websites that offer repetitive vocabulary practice are excellent ways for students to do this. Students may also use their notes, however, they will need to do repetitive activities, similar to flashcard practice, and not simply read and reread their notes.

➢ Bridging Math Application

Strategy 1: Prepare for math applications by identifying the problem type and the problem-solving strategies and tools.

Strategy 2: Organize the problem using visual, symbolic, and written representations.

Strategy 3: Overcome barriers to problem solving using math models, language and structural analysis, and resources.

Strategy 4: Demonstrate and defend problem-solving application and mathematical reasoning through reverse problem solving, mental mathematics, visual representations, and peer discussions.

1. Each part of this lesson lends itself to a variety of math application tasks that allow students to synthesize, apply, or extend their mathematical knowledge and skills. *Math Sense 1: Focus on Operations* Problem Solver Common Percent Applications (p. 130) offers a number of examples of real-life money applications for calculating percent. Use one of these as a springboard to create an authentic practice for students to apply their new knowledge. Or, here is another example not using money that provides an authentic practice for calculating percent.

> Good attendance in class is important for reaching our educational goals. Research studies show that the better your attendance is, the greater your progress. Furthermore, studies show that attendance under 60% often yields no progress at all. Is your attendance setting you up to succeed? How good is your attendance?
>
> Using the attendance data your teacher provides, calculate your percent of attendance for the past 4 weeks. Remember, 100% would be the total hours the class met in that time period. After you calculate your percent of attendance, evaluate how good your attendance is: Very Good (90–100%), Satisfactory (75–90%), Needs Improvement (60–75%), Unsatisfactory (below 60%). Now, calculate how many days you would need to come during the next 4 weeks to raise your attendance 10% and 20%. What can you do to make this happen? Write three action steps you can follow to improve your attendance by 10% to 20% over the next 4 weeks.

Before engaging in problem solving, have students analyze the directions to determine what is being asked of them and use this information to determine which strategies to use to complete the task. The following chart begins this process to provide a glimpse of how this may look.

Directions	Operation or Action *Signal Words*	Strategies and Tools
Using the attendance data your teacher provides, calculate your percent of attendance for the past 4 weeks. Remember, 100% would be the total hours the class met in that time period.	Calculate percent attendance	Total number of class hours = whole (100%) Use percent statement: percent of whole is part; Change to equation and put in known values and unknown variable: $n\% \times$ total class hours = my hours
After you calculate your percent of attendance, evaluate how good your attendance is: Very Good (90–100%), Satisfactory (75–90%), Needs Improvement (60–75%), Unsatisfactory (below 60%).		

2. Students will need to plan and organize how to complete the above task. Encourage students to create their own graphic organizer that outlines the steps they need to take and gives them a place to record their answers. This may take a lot of guidance to lead them toward an effective plan. For students who struggle with organizing, provide them a template to use but gradually release responsibility as they get more exposure to project planning.

Last 4 weeks Total Class Hours (whole)	**My Class Hours** (part)	**Percent Equation**	**My Attendance Percent** (unknown)
		$n\% \times w = p$ $n\% \times$ ______ $= p$	

Next 4 weeks Total Class Hours (whole)	**My Class Hours** (part-unknown)	**Percent Equation**	**My Attendance Goal** (Increase 10%)
		$n\% \times w = p$ ______% × ______ $= p$	My Percent + 10% = Goal

Next 4 weeks Total Class Hours (whole)	**My Class Hours** (part-unknown)	**Percent Equation**	**My Attendance Goal** (Increase 20%)
		$n\% \times w = p$ ______% × ______ $= p$	My Percent + 20% = Goal

3. Students can use the tools and strategies they have learned leading up to this task, such as the *Math Sense 1: Focus on Operations* lessons that feature step-by-step directions on how to create and solve percent equations.

4. To check their calculations, students can use reverse problem solving. In a percent equation, this can be done by using their answer as a known and changing one of the previously known variables to make it an unknown. For example, in the first calculation, the unknown is their percent of attendance. Students place the value they calculated for their percent of attendance into the equation and remove the known whole value replacing it with a variable. They solve the problem to see if they come up with the previously known whole. This can also be done, of course, by removing the known part.

➢ Assessment & Next Steps

Students should complete the practice activities included in each Math Sense 1: Focus on Operations lesson. Evaluate which learning goals were not met and remediate by using other resources, such as those identified in the Bridging Knowledge section. Upon successful completion, continue to the next unit.

Unit 5

MEASUREMENT

Skills-Based Questions:

1. What is the U.S. customary system and what are its units of length? How do you convert one customary unit of length to another? *(Part 1)*
2. How do you use the four operations with customary units of length? *(Part 2)*
3. How is capacity measured within the customary system? How do you convert capacity measurements? How do you use the four operations with customary units of capacity? *(Part 3)*
4. How is weight measured within the customary system? How do you convert weight measurements? How do you use the four operations with customary units of weight? *(Part 4)*
5. What is the organizing principle behind the metric system? How can you use this principle to convert measurements within the metric system? *(Part 5)*
6. How do you convert units of time? How do you find changes in temperature? *(Part 6)*

Math Sense 1: Focus on Operations: Part 1, p. 146; Part 2, p. 148; Part 3, p. 150; Part 4, p. 152; Part 5, p. 154; Part 6, p. 158

	Learning Goals:	**GED**
Knowledge Goals:	1. Describe the U.S. customary measurement system and its units of length. Explain how to convert between different units of length within the customary system. *(Part 1)*	Prep Q.4
	2. Describe how to add, subtract, multiply, and divide customary units of length. *(Part 2)*	
	3. Describe the customary units of capacity, how to convert between them, and how to add, subtract, multiply, and divide them. *(Part 3)*	
	4. Describe the customary units of weight, how to convert between them, and how to add, subtract, multiply, and divide them. *(Part 4)*	
	5. Describe how the metric system works, including how to convert its units of measurement. *(Part 5)*	
	6. Describe how to convert units of time and how to find the change in temperature. *(Part 6)*	
Problem-Solving Goals:	1. Convert between customary units of length. *(Part 1)*	Prep Q.4
	2. Add, subtract, multiply, and divide customary units of length. *(Part 2)*	
	3. Convert between customary units of capacity and add, subtract, multiply, and divide them. *(Part 3)*	
	4. Convert between customary units of weight and add, subtract, multiply, and divide them. *(Part 4)*	
	5. Convert between metric units of measurement. *(Part 5)*	
	6. Convert units between units of time and calculate change in temperature. *(Part 6)*	
Vocabulary Goals:	1. Define key mathematical terms.	
	2. Determine the meaning of unknown vocabulary using context clues, word forms, and parts of speech.	
	3. Apply new vocabulary to mathematical tasks and discussions.	
Math Application Goals:	1. Apply conceptual understanding of measurement to real life situations. *(Parts 1–6)*	Prep Q.4
	2. Defend math applications and reasoning to others. *(Parts 1–6)*	

Sample Instructional Support Strategies

➢ Bridging Knowledge

Strategy 1: ***Develop and connect background knowledge, skills, and conceptual understanding to new knowledge.***

Strategy 2: ***Use guiding questions to make connections beyond the lesson to broader math applications.***

Strategy 3: ***Use problem-solving strategies to develop, monitor, and synthesize conceptual understanding and fluency.*** *(See also Bridging Problem Solving)*

Strategy 4: ***Extend problem-solving skills and mathematical reasoning to broader math applications in life and work.*** *(See Bridging Math Application)*

1. Evaluate students' knowledge of the following mathematical skills. Utilize the chart below to develop student content knowledge as necessary.

<table>
<tr><th>Unit 5
Measurement</th><th>Part 1
Customary Units of Length</th><th>Part 2
Working with Length</th><th>Part 3
Measuring Capacity</th><th>Part 4
Measuring Weight</th><th>Part 5
Using Metric Units</th><th>Part 6
Working with Time and Temperature</th></tr>
<tr><td>Core Skills in Mathematics</td><td>Unit 6, Lesson 1: Measuring and Estimating Lengths in Standard Units (p. 111)</td><td colspan="5">Unit 6, Lesson 2: Solving Problems Involving Measurement (p. 115); Lesson 3: Converting Measurement Units (p. 119)</td></tr>
<tr><td rowspan="2">Scoreboost Mathematics: Measurement and Geometry</td><td colspan="4">Review the Standard Units of Measurement (p. 4)</td><td>Review the Metric Units of Measurement (p. 6)</td><td></td></tr>
<tr><td colspan="6">Use the Calculator w/ Measurement Problems (p. 8); Use Diagrams in Measurement Problems (p. 10); Measurement Strategies (p. 12)</td></tr>
</table>

2. The unit preview for each unit in the *Math Sense 1: Focus on Operations* provides a list of context examples in which to apply the target math concepts as well as questions that connect to students' prior knowledge and experience (Unit 5, p. 145). These questions also provide a bridge to broader math applications in life and work.

3. Each unit also provides specialized lessons that focus on problem solving, using tools, and test taking techniques using the math skills taught in the lesson. Utilize these lessons to build student knowledge in these areas.

Unit 5	Content	Page
Problem Solver	Making Conversions Figuring Distance, Rate, and Time Basic Perimeter Problems	156 160 162
Tools	NA	
Test Taker	Make Your Own Drawing	164

➢ Bridging Problem Solving

Strategy 1: ***Preview the problem to determine problem-solving strategies and tools and predict general solutions.***

Strategy 2: ***Develop conceptual understanding of mathematical problems using visual representations, think-alouds, and collaboration.***

Strategy 3: ***Overcome barriers to problem solving using math models, language and structural analysis, and resources.***

Strategy 4: ***Demonstrate and defend problem solving and mathematical reasoning through reverse problem solving, mental mathematics, visual representations, and peer discussions.***

1. For Parts 1–6 of Unit 5, orient students to features of each mathematical concept such as the symbols, language, and structure. Students should identify the math language and symbols to determine what is being asked of them and use this information to determine which strategies to use to complete the task. Although it is not necessary for students to fill out a graphic organizer for each math problem they attempt, completing the following graphic organizer is helpful for annotating and reviewing math concepts and choosing appropriate strategies to complete tasks. The following is an example of how a student (with guidance) might fill out this graphic organizer:

	Symbols Abbreviations	**Language**	**Operation or Action**	**Structure**	**Strategies/Tools**
Part 1 Customary: Length	′ ″ ft yd in mi	customary system, length, feet, yard, inch, mile, measurement, unit	measure, convert, multiply, divide	smaller to larger (fewer parts) = divide; larger to smaller (more parts)= multiply	set up multiplication problem moving from large (fewer) to small (more) parts; set up division problem moving from small (more) to large (fewer) parts
Part 2 Working with Length	+ − ÷ × =	unit, length, add, subtract, multiply, divide	add, subtract, multiply, divide	add/subtract: line up like units up and down; multiply/ divide each unit by the multiplier/divisor	set up addition and subtraction problem with like units up and down; set up multiplication and division problem in order to multiply/divide each unit by multiplier/divisor
Part 3 Customary: Capacity	+ − ÷ × = fl oz c pt qt gal	capacity, unit, measurement, liquid, granular, fluid ounce, cup, pint, quart, gallon	convert, multiply, divide, add, subtract	smaller to larger (fewer parts) = divide; larger to smaller (more parts) = multiply	convert: from large (fewer) to small (more) parts = multiply; from small (more) to large (fewer) parts = divide
Part 4 Customary: Weight	+ − ÷ × = oz lb t	weight, unit, ounce, pound, ton	convert, multiply, divide, add, subtract	smaller to larger (fewer parts) = divide; larger to smaller (more parts) = multiply	convert: from large (fewer) to small (more) parts = multiply; from small (more) to large (fewer) parts = divide
Part 5 Metric System	+ − ÷ × = . mm cm m km	metric system, unit, measurement, millimeter, centimeter, meter, kilometer, decimal point, place, tens	convert	smaller to larger (fewer parts) = decimal to left for each ten; larger to smaller (more parts) = decimal to right	use base ten to convert: for each base ten unit smaller, move one decimal place to the right (more parts); for each base ten unit larger, move one decimal place to the left (less parts)
Part 6 Time and Temperature	+ − ÷ × = : sec min hr d wk mo yr	time, second, minute, hour, day, week, month, year	add, subtract, multiply, divide	temperature change = large temp – smaller temp	use equivalencies to help (60, 24, 7, 12, 365); use knowledge of fractions to help when changing time decimals to minutes; subtract to find change of temperature

2. Measurement is about as hands-on as you can get when it comes to mathematics. Although measurement itself is not a GED target, knowing how to measure, convert between units of measurement, and use the four operations with measurements is essential for mastering geometry problems on the test as well as understanding a variety of other mathematical problems that use unit measurement within them. Providing students with the opportunity to measure using both the customary and metric systems helps prepare them for more difficult applications of these skills later on. Exploring equivalent measurements with different units helps build conceptual understanding of the process of converting. Here is one activity that provides students this opportunity.

Measurement Stations

Directions: In teams of 3 or 4, travel to each measurement station. Follow the directions at each station.

Station 1: Length
Materials: Ruler and measuring tape (two-sided: customary/metric)
Directions: Measure each item using each of the following units:

	Customary System		**Metric System**	
Length:	**Feet and Inches**	**Inches only**	**Centimeters**	**Millimeters**
Notebook length				
Table length				
Student height				
Etc...				

Team Discussion:

1. Look at the equivalent measurements. What relationships do you see between them?
2. Which measurement is larger/smaller? Feet or inches? Centimeters or millimeters? Inches or centimeters?
3. How can you change (convert) from one unit to another? Do you divide or multiply?
4. If you change from feet (fewer parts) to inches (more parts) do you divide or multiply? Inches to feet? Centimeters to millimeters? Millimeters to centimeters? Why?

Note: Create a chart and question set similar to this example for each additional station using capacity, weight, and time.

3. It is important to model each of these measurement tools before students use them. In addition, a guide that focuses in on the increments of each instrument will be helpful. For example, *Core Skills in Mathematics* Unit 6 Lessons 1–2 (pp. 111–118) provide excellent graphics of measurement tools to orient students to the increments of each. Images copied from the internet and put into a packet for students to refer to may also be helpful.

4. The team discussion in the above activity allows students to share their measurements and discuss their conceptual understanding. Students should take turns discussing the questions, suggesting answers, and probing the thinking of others. Accuracy is not the point of this activity; exploration is. Therefore, let students explore their ideas. Then use this activity to build in the mathematical skills of converting between units and adding, subtracting, multiplying, and dividing measurements.

➢ Bridging Vocabulary

Strategy 1: ***Identify the component parts and usage of new words to interpret their meanings.***

Strategy 2: ***Use context clues to interpret new words.***

Strategy 3: ***Utilize vocabulary-building resources.***

Strategy 4: ***Build a deeper knowledge of words through math application tasks and collaborative discussions.***

Strategy 5: ***Memorize words through repetitive study such as using flashcards (digital or print) and notes.***

1. First, present the shortest form of the word, referred to in this text as the "base word" in the case of academic words and some subject-specific terms. Follow the base form with other commonly used word forms (if available). Examine prefixes and suffixes and their impact on word meaning and part of speech.
2. Read the word as used in the context of the text and discuss possible meanings given context clues and word form.
3. Have students find (electronically or in print) the definition or translation of the base form and, if different, the form used in context and note these definitions in the space provided for future reference and study.
4. Gradually build a deeper knowledge of the word by having students use the word in a sentence frame, guided discussion, and an original sentence within a mathematical context.

Sentence Frame:	*The **capacity** of an object is* ________________ *or* ________________ *it* ________________.
Guided Discussion:	*What tools can you use to measure **capacity**? What units do we use to measure capacity?*
Original Math Sentence:	________________________________

Encourage students to use these words in math applications and collaborative discussions such as the task described in Bridging Problem Solving Strategy 4.

5. The high volume of mathematical terminology requires repeat exposure to the words over time. Word walls, intentionally including the words in questions to students and when eliciting responses from them, and explicit reminders to use the vocabulary in verbal tasks provide built-in reinforcement. However, this is often not enough so it is important that students learn ways to study words independently. Flashcards or websites that offer repetitive vocabulary practice are excellent ways for students to do this. Students may also use their notes, however, they will need to do repetitive activities, similar to flashcard practice, and not simply read and reread their notes.

➢ Bridging Math Application

Strategy 1: ***Prepare for math applications by identifying the problem type and the problem-solving strategies and tools.***

Strategy 2: ***Organize the problem using visual, symbolic, and written representations.***

Strategy 3: ***Overcome barriers to problem solving using math models, language and structural analysis, and resources.***

Strategy 4: ***Demonstrate and defend problem-solving application and mathematical reasoning through reverse problem solving, mental mathematics, visual representations, and peer discussions.***

1. Each part of this lesson lends itself to a variety of math application tasks that allow students to synthesize, apply, or extend their mathematical knowledge and skills. The following activity, gives students hands-on practice applying what they have learned during the unit about measurement and conversions.

Measurement Scavenger Hunt*

Teacher Directions:

Materials:

- Set of measuring cups and spoons
- Ruler and measuring tape
- Food scale
- Stopwatch
- Thermometer

Directions:

1. Divide students into groups of 3 or 4.
2. Give each group a bag with all the measuring tools they will need.
3. Give each group a list of items to measure and the unit to measure the item in.
4. Students work together to choose an appropriate measurement tool and measure each item.
5. Students record their measurements.
6. After students measure every item on their list, they will return to their seats and complete conversions to smaller and larger units. (These conversions should include both smaller and larger units from both the customary and metric systems.)
7. Students should complete their conversions and then share their results with another group.

*Don't have enough measurement instruments for each group? Set this activity up in stations and have each group rotate through the stations to complete the measurements and conversions.

Student Directions:

In a group, measure each item using the given unit of measurement and appropriate measurement tool. When you finish your measurements, convert them to the units listed in parentheses. When your group finishes its conversions, meet with another group to check your accuracy.

Measurement List:

1. The time it takes to finish this activity, in minutes and seconds (sec, hr, day)
2. The weight of a math book, in ounces (lb, g, mg)
3. The length of a bulletin board, in meters (in, ft/in, cm)
4. The capacity of a bag of rice, in grams (c, Tbsp, mg)
5. The weight of a ruler, in ounces (lb, g, mg)
6. The capacity of a container of water, in cups (fl oz, qt, l)
7. The time it takes to say the English alphabet, in seconds (min, hr, day)
8. The temperature of a glass of water, in degrees Fahrenheit (°C)
9. The height of another student, in feet and inches (in, m, cm)

Before engaging in problem solving, have students analyze the directions to determine what is being asked of them and use this information to determine which strategies to use to complete the task. The following chart begins this process to provide a glimpse of how this may look.

Directions	**Operation or Action** *Signal Words*	**Strategies and Tools**
In a group, measure each item using the given unit of measurement and appropriate measurement tool.	identify unit; choose tool, measure	Make a chart to fill in with each measurement. Label each measurement with the correct unit.
When you finish your measurements, convert them to the units listed in parentheses.	convert, multiply or divide	Identify the equivalent for each unit of measurement. Multiply to make more parts (smaller) or divide to make fewer parts (bigger). Use proportions when helpful.

2. Students will need to plan and organize how to complete the above task. Encourage students to create their own graphic organizer that outlines the steps they need to take and that gives them a place to record their answers. This may take a lot of guidance to lead them toward an effective plan. For students who struggle with organizing, provide them a template but gradually release responsibility as they get more exposure to project planning.

Measurement Scavenger Hunt

Materials:

- Set of measuring cups and spoons
- Ruler and measuring tape
- Food scale
- Stopwatch
- Thermometer

Directions:

1. Measure each item
2. Write measurement
3. Convert to each unit
4. Check accuracy with another group

Object-Unit	**Measurement**	**Conversion 1**	**Conversion 2**	**Conversion 3**
Activity-Time	_____ min:sec	_____ sec	_____ hr	_____ day
Math book-Weight	_____ oz	_____ lb	_____ g	_____ mg
Bulletin board-Length	_____ m	_____ in	_____ ft/in	_____ cm
Etc.				

3. Students can use the tools and strategies they have learned leading up to this task such as the *Math Sense 1: Focus on Operations* lessons that feature step-by-step directions on how to convert from one unit of measurement to another.

4. When students have completed the task and meet with another group to compare measurements and conversions, they will need to defend their conversions by explaining the mathematical process they used to derive them. Students should use precise mathematical language in their descriptions and engage in polite critique of others' calculations or processes when erroneous.

➢ Assessment & Next Steps

Students should complete the practice activities included in each Math Sense 1: Focus on Operations lesson. Evaluate which learning goals were not met and remediate by using other resources, such as those identified in the Bridging Knowledge section. Upon successful completion, continue to the Simulated GED Test.

Math Sense 2: Focus on Problem Solving

Unit 1

NUMBERS AND PROPERTIES

Skills-Based Questions:

1. What are signed numbers and how does a number line show the relationships between them? *(Part 1)*
2. What is an integer and how do you know if one integer is greater than, equal to, or less than another? *(Part 2)*
3. How do you add, subtract, multiply, and divide signed numbers? *(Parts 3 and 4)*
4. What are powers and square roots and how do you solve for them? *(Part 5)*
5. What is the order of operations and how is it helpful for finding the value of an expression? *(Part 6)*

Math Sense 2: Focus on Problem Solving: Part 1, p. 16; Part 2, p. 18; Part 3, p. 26; Part 4, p. 30; Part 5, p. 32; Part 6, p. 34

Learning Goals:

	Learning Goals	GED
Knowledge Goals:	1. Describe the relationship between signed numbers on a number line. *(Part 1)* 2. Describe the relationship (greater than, less than, equal to, the difference) between integers. *(Part 2)* 3. Explain how you add, subtract, multiply, and divide positive and negative numbers with the same signs and with different signs. *(Parts 3–4)* 4. Describe powers and square roots and explain how to solve for them. *(Part 5)* 5. Explain the steps in the order of operations and how to use them to find the value of a mathematical expression. *(Part 6)*	Q.1.b Q.1.c Q.1.d Q.2.a Q.2.b
Problem-Solving Goals:	1. Identify missing numbers on a number line. *(Part 1)* 2. Compare and order integers to determine which is greater or less than the other. *(Part 2)* 3. Add, subtract, multiply, and divide positive numbers together, negative numbers together, and positive and negative numbers together. *(Parts 3–4)* 4. Solve powers, especially squares and cubes, and square roots. *(Part 5)* 5. Use the order of operations to find the value of mathematical expressions. *(Part 6)*	Q.1.b Q.1.c Q.1.d Q.2.a Q.2.b Q.2.c
Vocabulary Goals:	1. Define key mathematical terms. 2. Determine the meaning of unknown vocabulary using context clues, word forms, and parts of speech. 3. Apply new vocabulary to mathematical tasks and discussions.	
Math Application Goals:	1. Apply knowledge of signed numbers, exponents, square roots, number properties and order of operations to solve real-life math problems. *(Parts 1–6)* 2. Defend math applications and reasoning to others. *(Parts 1–6)*	Q.1.b Q.1.d Q.2.a Q.2.b Q.2.c

Sample Instructional Support Strategies

➤ Bridging Knowledge

Strategy 1: ***Develop and connect background knowledge, skills, and conceptual understanding to new knowledge.***

Strategy 2: ***Use guiding questions to make connections beyond the lesson to broader math applications.***

Strategy 3: ***Use problem-solving strategies to develop, monitor, and synthesize conceptual understanding and fluency.*** *(See also Bridging Problem Solving)*

Strategy 4: ***Extend problem-solving skills and mathematical reasoning to broader math applications in life and work.*** *(See Bridging Math Application)*

1. Evaluate students' knowledge of the following mathematical skills. Utilize the chart below to develop student content knowledge as necessary.

Unit 1 **Numbers and Properties**	**Part 1** The Number Line	**Part 2** Comparing and Ordering Integers	**Part 3** Adding and Subtracting Signed Numbers	**Part 4** Multiplying and Dividing Signed Numbers	**Part 5** Powers and Roots	**Part 6** Order of Operations
Core Skills in Mathematics	Unit 1, Lesson 1: Presenting Numbers on a Number Line (p. 12)				Unit 4, Lesson 1: Evaluating Expressions (p. 60)	
Scoreboost Mathematics: Algebraic Reasoning			Add and Subtract Integers (p. 4)	Multiply and Divide Integers (p. 6)	Find Powers and Roots (p. 8)	Evaluate Expressions (p. 10)
Pre-HSE Workbook: Math 2						Using Algebraic Expressions and Variables (p. 16)

2. The unit preview for each unit in *Math Sense 2: Focus on Problem Solving* provides a list of context examples in which to apply the target math concepts as well as questions that connect to students' prior knowledge and experience (for Unit 1, p. 15). These questions also provide a bridge to broader math applications in life and work.

3. Each unit also provides specialized lessons that focus on problem solving, using tools, and test taking techniques using the math skills taught in the lesson. Utilize these lessons to build student knowledge in these areas.

Unit 1	**Content (GED Target)**	**Page**
Problem Solver	Kinds of Numbers Prime Numbers and Prime Factoring (Q.1.b) Absolute Value (Q.1.d)	20 22 28
Tools	Properties of Numbers Using a Calculator	24 36
Test Taker	NA	

➢ Bridging Problem Solving

Strategy 1: Preview the problem to determine problem-solving strategies and tools and predict general solutions.

Strategy 2: Develop conceptual understanding of mathematical problems using visual representations, think-alouds, and collaboration.

Strategy 3: Overcome barriers to problem solving using math models, language and structural analysis, and resources.

Strategy 4: Demonstrate and defend problem solving and mathematical reasoning through reverse problem solving, mental mathematics, visual representations, and peer discussions.

1. For Parts 1–6 of this unit, orient students to features of each mathematical concept such as the symbols, language, and structure. Students should identify the math language and symbols to determine what is being asked of them and use this information to determine which strategies to use to complete the task. Although it is not necessary for students to fill out a graphic organizer for each math problem they attempt, completing the following graphic organizer is helpful for annotating and reviewing math concepts and choosing appropriate strategies to complete tasks. The following is an example of how a student (with guidance) might fill out this graphic organizer:

	Symbols	Language	Operation or Action	Structure	Strategies/Tools
Part 1 Number Line	+ – ( )	whole numbers, fractions, decimals, signed numbers, positive, negative	identify	numbers to the right are larger than zero = positive; numbers to left are smaller than zero = negative	use the number line pattern to identify missing numbers
Part 2 Compare Integers	+ – < > =	integer, greater than, less than, equal to, positive, negative	compare, find	move right: numbers become greater; move left: numbers become smaller	use the number line to compare integers
Part 3 Add and Subtract Signed Numbers	+ – = ( )	number line, add, subtract, signed numbers, positive, negative	add, subtract	if signs are the same, add and keep sign; if signs are different, subtract and keep sign of larger number	move from 1 point to the next on the number line to add and subtract numbers
Part 4 Multiply and Divide Signed Numbers	× ÷ $\overline{)\ }$	multiply, divide, signed numbers, positive, negative	multiply, divide	if signs are the same, positive answer; if signs are different, negative answer	use rules for multiplying and dividing signed numbers
Part 5 Powers and Roots	5^2 $\sqrt{\ }$	power, exponent, base, root, square, cube, radical sign, perfect square	solve	multiply base by itself as many times as the exponent	think of a square for exponents of 2 and cubes for exponents of 3; use perfect squares to find approximate square root
Part 6 Order of Operations	+ – × ÷ ( ) { } $\overline{)\ }$ $\sqrt{\ }$	order of operations, expression, value, grouping, parentheses, brackets, fraction bar, radical sign, powers, exponents, roots,	find, perform, evaluate, multiply, divide, add, subtract	P=Parentheses E=Exponents and Roots M/D= Multiply and Divide A/S= Add and Subtract	Follow PEMDAS to find the value of expressions

2. Expanding algebraic thinking requires further exploration with math concepts in concrete ways. The number line is one concrete way to build conceptual understanding of signed numbers. Here are other suggestions for activities that allow students to explore the concepts in Parts 1–6 of this unit.

Parts 1–3: Comparing and subtracting signed numbers	Chart the average seasonal temperatures of Alaska on a drawing of a thermometer. Compare the temperatures from season to season and find the difference between them. Or, do the same thing on a floor thermometer and have students walk the difference in temperatures.
Parts 3–4: Adding, multiplying, and dividing signed numbers	**Preparation:** Prepare a checking account balance statement for each student in the class. (This may be on a note card or half-sheet of paper.) Prepare balances that are both positive and negative. Also, prepare several bank balances that are the same amount. [Example: Student 1 Balance = $250.00; Student 2 Balance = -$45.00; Student 3 Balance = $110.00; Student 4 Balance = -$45.00; Student 5 Balance = $250.00] **Addition Activity:** Give each student one of the prepared checking account balance statements. In groups of three to five students have them add up the total balance for their group. Then have groups share the individual balances as well as their group's total, which the teacher writes on the board. Once all groups have shared, add up all the totals to find the total balance for the class. [Example: $250 + $250 + -$45 + -$45 + $110 = total ($520)] **Multiplication Activity:** Now, take advantage of the duplicate bank balances and go back to see if you can re-group the individual student balances in order to apply multiplication to recalculate the total balance for the class. [Example: (2)($250) + (2)(-$45) + $110 = total] **Division Activity:** Divide the total class balance by the number of group balances to find the average balance.
Part 5: Powers and roots	Use base ten blocks to explore powers and roots. Using the single blocks, students can explore solving powers (especially squares and cubes) by connecting blocks, and explore roots by separating them.
Part 6: Order of operations	Use base ten blocks in their three forms (single blocks, sticks of 10, and tiles of 100) to explore the order of operations. Give students simple equations to solve by mixing up the order of operations and by using the order. Students may find that they can still get the right answer by changing the order but that using the order will always give them the correct answer.

3. Journeying into algebraic thinking starts the journey into the depths of math language. Providing cheat sheets with math vocabulary definitions and examples may help students overcome the roadblock that the precise algebraic language produces. The chart from Strategy 1 is a good place to start. The glossary in *Math Sense 2: Focus on Problem Solving* is another resource. It is essential that students develop a working knowledge of this vocabulary to make way for accurate use of the concepts and skills behind it.

4. Each of the exploration activities suggested in Strategy 2 should be accompanied by the opportunity for students to share the reasoning behind the concepts they are building. In other words, students need to come together and explain, using precise mathematical terminology, their new understandings of each math concept they have explored. The Skills-Based Questions from the beginning of this lesson can guide this exchange. After the exploration activity, set up students in pairs or small groups and have them discuss the question (or questions) that pertain to the concept they just explored. Students should also include examples from the exploration activity in their explanations. You may want students to summarize their understanding by writing an explanation or creating a visual that illustrates the answer.

➢ Bridging Vocabulary

Strategy 1: Identify the component parts and usage of new words to interpret their meanings.

Strategy 2: Use context clues to interpret new words.

Strategy 3: Utilize vocabulary-building resources.

Strategy 4: Build a deeper knowledge of words through math application tasks and collaborative discussions.

Strategy 5: Memorize words through repetitive study such as using flashcards (digital or print) and notes.

1. First, present the shortest form of the word, referred to in this text as the "base word" in the case of academic words and some subject-specific terms. Follow the base form with other commonly used word forms (if available). Examine prefixes and suffixes and their impact on word meaning and part of speech.
2. Read the word as used in the context of the text and discuss possible meanings given context clues and word form.
3. Have students find (electronically or in print) the definition or translation of the base form and, if different, the form used in context and note these definitions in the space provided for future reference and study.
4. Gradually build a deeper knowledge of the word by having students use the word in a sentence frame, guided discussion, and an original sentence within a mathematical context.

Sentence Frame:	*To solve a* ***power****, you must* ______ *the* ______ *by itself as many times as the* ______.
Guided Discussion:	*Why are* ***powers*** *useful? Where do we use them in real life?*
Original Math Sentence:	______

Encourage students to use these words in math applications and collaborative discussions such as the task described in Bridging Problem Solving Strategy 4.

5. The high volume of mathematical terminology requires repeat exposure to the words over time. Word walls, intentionally including the words in questions to students and when eliciting responses from them, and explicit reminders to use the vocabulary in verbal tasks provide built-in reinforcement. However, this is often not enough so it is important that students learn ways to study words independently. Flashcards or websites that offer repetitive vocabulary practice are excellent ways for students to do this. Students may also use their notes, however, they will need to do repetitive activities, similar to flashcard practice, and not simply read and reread their notes.

➢ Bridging Math Application

Strategy 1: ***Prepare for math applications by identifying the problem type and the problem-solving strategies and tools.***

Strategy 2: ***Organize the problem using visual, symbolic, and written representations.***

Strategy 3: ***Overcome barriers to problem solving using math models, language and structural analysis, and resources.***

Strategy 4: ***Demonstrate and defend problem-solving application and mathematical reasoning through reverse problem solving, mental mathematics, visual representations, and peer discussions.***

1. Each part of this lesson lends itself to a variety of math application tasks that allow students to synthesize, apply, or extend their mathematical knowledge and skills. Whichever math application task you choose, be sure to orient students to the problem type and the problem-solving strategies and tools they may utilize. The following is an example of student directions for a math application task that synthesizes the concepts developed in this lesson.

Redistributing Wealth

Your teacher will give you and each of your classmates a statement of your total net worth (the total of all the money and possessions you have, or your wealth). This distribution of net worth will be similar to the distribution of wealth in the United States*. Your class has decided to redistribute the total wealth so that each student has an equal net worth. First, decide how much the total net worth of the class is. Next, determine how to redistribute the wealth so that each person gets the same amount. Finally, answer the following questions using your calculation.

How much money did each of the poorest people in class gain?

How much did they gain in total?

How much money did the richest person give?

The population of the United States is approximately 3.2^8 people. If the richest person in your class equals 1% of this population, how much wealth would the richest 1 percent of the people in the United States give?

*Note to Teacher: Use reliable and current statistics for wealth distribution in the United States to create the net worth amounts for each student.

Before engaging in problem solving, have students analyze the directions to determine what is being asked of them and use this information to determine which strategies to use to complete the task. The following example provides a glimpse of how this may look.

Directions	**Operation or Action** *Signal Words*	**Strategies and Tools**
First, decide how much the total net worth of the class is.	addition	Add all positive numbers; add all negative numbers; add together and keep the sign of the largest number.
Next, determine how to redistribute the wealth so that each person gets the same amount.		

2. Students will need to plan and organize how to complete the above task. Encourage students to create their own graphic organizer that outlines the steps they need to take and gives them a place to record their answers. This may take a lot of guidance to lead them toward an effective plan. For students who struggle with organizing, provide them a template but gradually release responsibility as they get more exposure to project planning.

Step	Calculation	Answer

3. Working with peers is a good way for students to develop and hone their mathematical knowledge and skills through discussions and tasks that require them to explain (and sometimes defend) their own mathematical reasoning and probe that of others. Using the Bridging Math Application example, students work in small groups to organize and accomplish the task. During this collaborative work, students can practice both precise mathematical terms to describe their ideas and mathematical reasoning as well as the discourse prompts (p. 159) necessary to work well in a group.

Polite disagreement: *I see what you're saying but I think ________. Another way to think about this is ________.*

Encourage participation: *What do you think? I'd like to hear what you have to say about ________.*

Probe others' ideas: *Could you explain that further? What did you mean by ________?*

Paraphrase: *So, do we agree that ________? I think you're saying that ________.*

Present examples: *An example of this is ________. ________ is an example of ________.*

4. As mentioned in Strategy 3, collaboration with students, including student discussions, helps students overcome barriers to problem solving but it also makes them accountable for their own mathematical skills and reasoning. This can be accomplished through a jigsaw activity. Have each group pair up with another group. Each group takes a turn sharing their calculations. Stress to students the importance of using precise mathematical language in their explanations and supporting them with sound mathematical reasoning. A good way to set up this exchange is to provide a checklist of student expectations.

☑ Take turns with your group members in describing each step of the task.

☑ Take turns explaining the mathematical reasoning behind each of the calculations.

☑ Use precise mathematical language.

➢ Assessment & Next Steps

Students should complete the practice activities included in each *Math Sense 2: Focus on Problem Solving* lesson. Evaluate which learning goals were not met and remediate by using other resources, such as those identified in the Bridging Knowledge section. Upon successful completion, continue to the next unit.

Unit 2

THE BASICS OF ALGEBRA: Section 1

Skills-Based Questions:

1. What are three mathematical statements used in algebra? How do you translate their symbols into words? *(Part 1)*
2. What does it mean to evaluate an expression? How do you evaluate expressions? *(Part 2)*
3. How do you simplify expressions? *(Part 3)*
4. How do you evaluate expressions with negative exponents? *(Part 4)*
5. What are radicals and how do you simplify them in expressions? *(Part 5)*

Math Sense 2: Focus on Problem Solving: Part 1, p. 42; Part 2, p. 44; Part 3, p. 46; Part 4, p. 50; Part 5, p. 54

	Learning Goals:	GED
Knowledge Goals:	1. Describe, in words and symbols, different ways to show the relationship between numbers and variables. *(Part 1)* 2. Describe how to evaluate an expression. *(Part 2)* 3. Explain how to simplify expressions. *(Part 3)* 4. Describe how to evaluate expressions with negative exponents. *(Part 4)* 5. Explain how to simplify radicals in expressions. *(Part 5)*	A.1.e A.1.g A.1.j A.1.i Q.1.c
Problem-Solving Goals:	1. Translate expressions, equations, and inequalities from symbols to words and from words to symbols. *(Part 1)* 2. Evaluate expressions by replacing variables with given values. *(Part 2)* 3. Simplify expressions by grouping like terms and numbers. *(Part 3)* 4. Evaluate expressions with negative exponents. *(Part 4)* 5. Use properties of square roots to simplify radicals. *(Part 5)*	A.1.e A.1.g A.1.j A.1.i Q.1.c
Vocabulary Goals:	1. Define key mathematical terms. 2. Determine the meaning of unknown vocabulary using context clues, word forms, and parts of speech. 3. Apply new vocabulary to mathematical tasks and discussions.	
Math Application Goals:	1. Apply knowledge of the basics of algebra to solve real-life math problems. *(Parts 1–5)* 2. Defend math applications and reasoning to others. *(Parts 1–5)*	A.1.e A.1.g A.1.j A.1.i Q.1.c

Sample Instructional Support Strategies

➢ Bridging Knowledge

Strategy 1: ***Develop and connect background knowledge, skills, and conceptual understanding to new knowledge.***

Strategy 2: ***Use guiding questions to make connections beyond the lesson to broader math applications.***

Strategy 3: ***Use problem-solving strategies to develop, monitor, and synthesize conceptual understanding and fluency.*** *(See also Bridging Problem Solving)*

Strategy 4: ***Extend problem-solving skills and mathematical reasoning to broader math applications in life and work.*** *(See Bridging Math Application)*

1. Evaluate students' knowledge of the following mathematical skills. Utilize the chart below to develop student content knowledge as necessary.

Unit 2 **The Basics of Algebra, Section 1**	**Part 1** Expressions and Variables	**Part 2** Evaluating Expressions	**Part 3** Simplifying Expressions	**Part 4** Negative Exponents	**Part 5** Simplifying Radicals
Core Skills in Mathematics	Unit 4, Lesson 1: Evaluating Expressions (p. 60)				
Scoreboost Mathematics: Algebraic Reasoning	Evaluate Expressions (p. 10)			Find Powers and Roots (p. 8)	
Pre-HSE Workbook: Math 2	Using Algebraic Expressions and Variables (p. 16)				

2. The unit preview for each unit in the *Math Sense 2: Focus on Problem Solving* provides a list of context examples in which to apply the target math concepts as well as questions that connect to students' prior knowledge and experience (for Unit 2, p. 41). These questions also provide a bridge to broader math applications in life and work.

3. Each unit also provides specialized lessons that focus on problem solving, using tools, and test taking techniques using the math skills taught in the lesson. Utilize these lessons to build student knowledge in these areas.

Unit 2, Section 1	Content (GED Target)	Page
Problem Solver	NA	
Tools	The Rules of Exponents (Q.1.c) Scientific Notation	48 52
Test Taker	NA	

➢ Bridging Problem Solving

Strategy 1: ***Preview the problem to determine problem-solving strategies and tools and predict general solutions.***

Strategy 2: ***Develop conceptual understanding of mathematical problems using visual representations, think-alouds, and collaboration.***

Strategy 3: ***Overcome barriers to problem solving using math models, language and structural analysis, and resources.***

Strategy 4: ***Demonstrate and defend problem solving and mathematical reasoning through reverse problem solving, mental mathematics, visual representations, and peer discussions.***

1. For Parts 1–5 of Unit 2, orient students to features of each mathematical concept such as the symbols, language, and structure. Students should identify the math language and symbols to determine what is being asked of them and use this information to determine which strategies to use to complete the task. Although it is not necessary for students to fill out a graphic organizer for each math problem they attempt, completing the following graphic organizer is helpful for annotating and reviewing math concepts and choosing appropriate strategies to complete tasks. The following is an example of how a student (with guidance) might fill out this graphic organizer:

	Symbols	Language	Operation or Action	Structure	Strategies/Tools
Part 1 Expressions and Variables	x n + – =	variable, expression, equation, inequality, product, quotient, difference, times, sum, less than, greater than, divided by, increased by, decreased by	translate, write	expressions show relationship between numbers and variables using operations; equations show what expressions equal	match operations to words
Part 2 Evaluate Expressions	x n + – = ( )	expression, variable, value, order of operations, negative, negation symbol, grouping symbol	evaluate, find, negate, group	do operations in grouping symbols ( ) | | first; negation symbol means it is opposite of the shown value	use order of operations (PEMDAS)
Part 3 Simplify Expressions	x n + – = ( )	equivalent expressions, variables, simplified, like terms, coefficients	simplify, group	group like terms and then numbers	apply rules for using operations with signed numbers
Part 4 Negative Exponents	5^{-2} n^{-5} – $\frac{\blacksquare}{\blacksquare}$	negative exponents, canceling, fraction bar, numerator, denominator, variables, given values	solve, cancel, find, substitute	change a negative exponent to a positive by moving the base and exponent to the other side of the fraction bar	follow rules of exponents; cancel out powers with like bases
Part 5 Simplify Radicals	$\sqrt{\ }$ =	simplify, radicals, square root, perfect square, product property, quotient property, factor	find, solve, estimate, simplify, factor	product's square root = the square root of each factor multiplied together; quotient's square root = radical numerator and radical denominator	follow the properties of square roots, product and quotient rules

2. Expanding algebraic thinking requires further exploration with math concepts in concrete ways. Here are some suggestions for activities that allow students to explore the concepts in each part of this section of Unit 2.

Parts 1–3: Evaluating Expressions	Divide students into groups of three or four. Give each group a different simple algebraic expression (for example, $a + 2b$). Tell them what the variables represent (for example, weight). Students can then choose what they want the coefficients to represent (people, apples, elephants, etc.). Students can then connect the expression to real life by drawing the coefficients within the expression: a + b. Next, they replace variables a and b with values of their choosing: (1,000 lbs) + (3,000 lbs). Finally, students can find the value of the expression (in this case, 7,000 lbs). Repeat this several times with different algebraic expressions that allow them to explore different operations and scenarios.
Parts 4–5: Using Rules of Exponents and Square Roots	One great way to get a job is through networking. Have students discuss networking through social media. How does it work? How can someone reach people they don't even know? Tell students that they are (hypothetically) going to use social media to find a job. On Day 1, they ask their 3 closest friends if they know of any job openings. On Day 2, those friends ask 3 of their friends. On Day 3, each of those friends ask 3 more friends and so on until a week has passed. In one week, how many people will be helping you find a job? What if you asked 5 friends and they each asked 5 and so on for one week?

3. Journeying into algebraic thinking starts the journey into the depths of math language. Providing cheat sheets with math vocabulary definitions and examples may help students overcome the roadblock that the precise algebraic language produces. The chart from Strategy 1 is a good place to start. The glossary in *Math Sense 2: Focus on Problem Solving* is another resource. It is essential that students develop a working knowledge of this vocabulary to make way for accurate use of the concepts and skills behind it.

4. Once students find the value of the algebraic expressions they explored in the activity from Strategy 2, they can create algebraic word problems to accompany them. They can also write a word problem for a different group's illustrated expression. For example:

   A circus has 3 elephants. One elephant is 1,000 pounds. Two elephants are 3,000 pounds each. The circus master must load all the elephants into a big trailer. How many pounds must the trailer be able to carry?

   Then, students can take turns solving the word problems from other groups. Not all illustrated expressions or word problems will make sense. Therefore, students will have to probe the reasoning of others as well as explain their own.

➢ Bridging Vocabulary

Strategy 1: Identify the component parts and usage of new words to interpret their meanings.

Strategy 2: Use context clues to interpret new words.

Strategy 3: Utilize vocabulary-building resources.

Strategy 4: Build a deeper knowledge of words through math application tasks and collaborative discussions.

Strategy 5: Memorize words through repetitive study such as using flashcards (digital or print) and notes.

1. First, present the shortest form of the word, referred to in this text as the "base word" in the case of academic words and some subject-specific terms. Follow the base form with other commonly used word forms (if available). Examine prefixes and suffixes and their impact on word meaning and part of speech.
2. Read the word as used in the context of the text and discuss possible meanings given context clues and word form.
3. Have students find (electronically or in print) the definition or translation of the base form and, if different, the form used in context and note these definitions in the space provided for future reference and study.

4. Gradually build a deeper knowledge of the word by having students use the word in a sentence frame, guided discussion, and an original sentence within a mathematical context.

Sentence Frame:	*I use a **variable** in an expression when ________________.*
Guided Discussion:	*Why are **variables** useful in math? Why are they useful in our lives?*
Original Math Sentence:	______________________________

Encourage students to use these words in math applications and collaborative discussions such as the task described in Bridging Problem Solving Strategy 4.

5. The high volume of mathematical terminology requires repeat exposure to the words over time. Word walls, intentionally including the words in questions to students and when eliciting responses from them, and explicit reminders to use the vocabulary in verbal tasks provide built-in reinforcement. However, this is often not enough so it is important that students learn ways to study words independently. Flashcards or websites that offer repetitive vocabulary practice are excellent ways for students to do this. Students may also use their notes, however, they will need to do repetitive activities, similar to flashcard practice, and not simply read and reread their notes.

➤ Bridging Math Application

Strategy 1: Prepare for math applications by identifying the problem type and the problem-solving strategies and tools.

Strategy 2: Organize the problem using visual, symbolic, and written representations.

Strategy 3: Overcome barriers to problem solving using math models, language and structural analysis, and resources.

Strategy 4: Demonstrate and defend problem-solving application and mathematical reasoning through reverse problem solving, mental mathematics, visual representations, and peer discussions.

1. Each part of this lesson lends itself to a variety of math application tasks that allow students to synthesize, apply, or extend their mathematical knowledge and skills. Whichever math application task you choose, be sure to orient students to the problem type and the problem-solving strategies and tools they may utilize. The following is an example of student directions for a math application task that synthesizes the concepts developed in this lesson.

Fruit Salad

Part 1:
In a small group, read the recipe for fruit salad. You need to make the recipe for 20 people. Determine how much of each ingredient you need for the salad. Although you do not know the cost of each item yet, create an algebraic expression, using variables for the missing costs. Once you have your expression ready, meet with another group and compare expressions. When both of your groups agree that the expression makes sense, ask the teacher for a list of the costs of each item. Use the list of costs to replace each variable with each item's cost. What is the value of your expression? How much will the ingredients cost?

Fruit Salad (Serves 4 people)

1 apple

1 banana

½ lb. grapes

⅛ lb. strawberries

1 8-oz. jar of marshmallow creme

Part 2:
Your fruit salad was so amazing that each of the 20 people who ate the salad told 4 other people how good it was. The next day, each of these people told 4 others and so on with new person telling 4 other people. After 5 days, how many people know how delicious your fruit salad was? How much fruit would you have to buy if you had to make fruit salad for all of these people? How much would the total ingredients cost? Create algebraic expressions to answer each of these questions. Use the list of costs of each food item for your calculations.

Before engaging in problem solving, have students analyze the directions to determine what is being asked of them and use this information to determine which strategies to use to complete the task. The following example provides a glimpse of how this may look.

Directions	**Operation or Action** *Signal Words*	**Strategies and Tools**
You need to make the recipe for 20 people. Determine how much of each ingredient you need for the salad.	multiply	Set up the problem. Multiply each ingredient by 20. Simplify fractions as needed.
Create an algebraic expression, using variables for the missing costs.		

2. Students will need to plan and organize how to complete the above task. Encourage students to create their own graphic organizer that outlines the steps they need to take and gives them a place to record their answers. This may take a lot of guidance to lead them toward an effective plan. For students who struggle with organizing, provide them a template but gradually release responsibility as they get more exposure to project planning.
3. Strategy 1 above gives students the opportunity to analyze the language and the structure of the task. This is especially beneficial for ELLs, so make sure they have plenty of time to dissect the problem into its component steps, grasp the meaning of the language being used, and determine what operations and/or actions are required of them.
4. Once students determine what they need to do for the problem (and write it in the Operation or Action column), have them check with their peers to see if their reasoning makes sense. A simple way to share one's mathematical reasoning is to determine: what the problem says; what I think that means we need to do; and why I think it means that.

The problem tells us to ______________ *and I think that means that we need to* ______________ *because* ______________.

➢ Assessment & Next Steps

Students should complete the practice activities included in each *Math Sense 2: Focus on Problem Solving* lesson. Evaluate which learning goals were not met and remediate by using other resources, such as those identified in the Bridging Knowledge section. Upon successful completion, continue to the next section of the unit.

THE BASICS OF ALGEBRA: Section 2

Skills-Based Questions:

1. How do you solve addition and subtraction equations? What are inverse operations and how do they work? *(Part 6)*
2. How do you solve multiplication and division equations? *(Part 7)*
3. How do you solve multistep equations that use any of the four operations? *(Part 8)*
4. What is a term? How do you solve equations that have separated terms? *(Part 9)*
5. What do parentheses ( ) mean in an equation? How do you solve equations with parentheses? *(Part 10)*
6. What is an inequality? How do you solve for inequalities? How do you graph them? *(Part 11)*

Math Sense 2: Focus on Problem Solving: Part 6, p. 60; Part 7, p. 62; Part 8, p. 64; Part 9, p. 66; Part 10, p. 68; Part 11, p. 72

	Learning Goals:	**GED**
Knowledge Goals:	1. Describe how you use inverse operations to solve addition and subtraction equations. *(Part 6)*	A.1.d
	2. Describe how to use inverse operations to solve multiplication and division equations. *(Part 7)*	A.1.h
	3. Explain how to solve multistep equations that use any of the four operations. *(Part 8)*	A.3.b
	4. Describe how to combine like terms to solve equations with separated terms *(Part 9)*	
	5. Explain the use of parentheses in an equation and how to solve an equation that uses them. *(Part 10)*	
	6. Describe what inequalities are and how to solve and graph them. *(Part 11)*	
Problem-Solving Goals:	1. Solve addition and subtraction equations using inverse operations. *(Part 6)*	A.1.d
	2. Solve addition and subtraction equations using inverse operations. *(Part 7)*	A.1.h
	3. Solve multistep equations using the four operations. *(Part 8)*	A.3.b
	4. Combine like terms to solve equations with separated terms. *(Part 9)*	
	5. Solve equations that use parentheses and other grouping symbols. *(Part 10)*	
	6. Solve and graph inequalities. *(Part 11)*	
Vocabulary Goals:	1. Define key mathematical terms.	
	2. Determine the meaning of unknown vocabulary using context clues, word forms, and parts of speech.	
	3. Apply new vocabulary to mathematical tasks and discussions.	
Math Application Goals:	1. Apply knowledge of algebraic equations and inequalities to solve real-life math problems. *(Parts 6–11)*	A.1.d
		A.1.h
	2. Defend math applications and reasoning to others. *(Parts 6–11)*	A.3.b

Sample Instructional Support Strategies

➢ Bridging Knowledge

Strategy 1: ***Develop and connect background knowledge, skills, and conceptual understanding to new knowledge.***

Strategy 2: ***Use guiding questions to make connections beyond the lesson to broader math applications.***

Strategy 3: ***Use problem-solving strategies to develop, monitor, and synthesize conceptual understanding and fluency.*** *(See also Bridging Problem Solving)*

Strategy 4: ***Extend problem-solving skills and mathematical reasoning to broader math applications in life and work.*** *(See Bridging Math Application)*

1. Evaluate students' knowledge of the following mathematical skills. Utilize the chart below to develop student content knowledge as necessary.

Unit 2 **The Basics of Algebra, Section 2**	**Part 6** Solving Addition and Subtraction Equations	**Part 7** Solving Multiplication and Division Equations	**Part 8** Solving Multistep Equations	**Part 9** Solving Equations with Separated Terms	**Part 10** Solving Equations with Parentheses	**Part 11** Solving Inequalities
Core Skills in Mathematics	Unit 4, Lesson 2: Solving Equations (p. 64); Lesson 5: Solving Problems Using Expressions, Equations, and Inequalities (p. 76)					Unit 4, Lesson 4: Graphing Equations and Inequalities (p. 72)
Scoreboost Mathematics: Algebraic Reasoning	Write and Solve Equations (p. 14)					Solve Inequalities (p. 18)
Pre-HSE Workbook: Math 2	Solving Equations (p. 18)		Solving Multistep Equations (p. 20)			Solving Inequalities (p. 28)

2. The unit preview for each unit in the *Math Sense 2: Focus on Problem Solving* provides a list of context examples in which to apply the target math concepts as well as questions that connect to students' prior knowledge and experience (for Unit 2, p. 41). These questions also provide a bridge to broader math applications in life and work.

3. Each unit also provides focus lessons on problem solving, using tools, and test taking. Use these lessons to bridge student knowledge in these areas.

Unit 2, Section 2	Content (GED Target)	Page
Problem Solver	Reading and Writing Equations (A.1.c, A.1.g, A.1.j)	58
Tools	Graphing Inequalities (A.3.b)	70
Test Taker	Try the Answer Choices	74

➤ Bridging Problem Solving

Strategy 1: ***Preview the problem to determine problem-solving strategies and tools and predict general solutions.***

Strategy 2: ***Develop conceptual understanding of mathematical problems using visual representations, think-alouds, and collaboration.***

Strategy 3: ***Overcome barriers to problem solving using math models, language and structural analysis, and resources.***

Strategy 4: ***Demonstrate and defend problem solving and mathematical reasoning through reverse problem solving, mental mathematics, visual representations, and peer discussions.***

1. For Parts 6–11 of Unit 2, orient students to features of each mathematical concept such as the symbols, language, and structure. Students should identify the math language and symbols to determine what is being asked of them and use this information to determine which strategies to use to complete the task. Although it is not necessary for students to fill out a graphic organizer for each math problem they attempt, completing the following graphic organizer is helpful for annotating and reviewing math concepts and choosing appropriate strategies to complete tasks. The following is an example of how a student (with guidance) might fill out this graphic organizer:

	Symbols	Language	Operation or Action	Structure	Strategies/Tools
Part 6 Addition and Subtraction Equations	x n + − =	addition, subtraction, equation, variable, inverse operations, simplify, numerical terms	add, subtract, simplify, combine, solve	addition and subtraction are inverse operations; equation is equal on each side	use inverse operations to remove something from one side; use the same operation to move it to the other side
Part 7 Multiplication and Division Equations	× ÷ ( ) − =	multiplication, division, equation, variable, inverse operations, reciprocal, numerator, denominator	multiply, divide, solve,	multiplication and division are inverse operations; multiply a fraction by its reciprocal to get 1	reverse the order of operations: SA, DM, E, P
Part 8 Multistep Equations	+ − × ÷ ( ) $\frac{\blacksquare}{\blacksquare}$ =	addition, subtraction, multiplication, division, inverse operations, variable	add, subtract, multiply, divide, solve		apply rules for using operations with signed numbers
Part 9 Separated Terms	+ − × ÷ ( ) $\frac{\blacksquare}{\blacksquare}$ = n	addition, subtraction, multiplication, division, inverse operations, variable, term, like terms	add, subtract, multiply, divide, combine	like terms contain the same variable; use the given operations to combine them before solving	simplify the equation by combining like terms; isolate the variable using inverse operations
Part 10 Equations with Parentheses	+ − × ÷ ( ) $\frac{\blacksquare}{\blacksquare}$ = n	parentheses, grouping symbols, distributive property, fraction bar	separate, combine, solve, add, subtract, multiply, divide	parentheses with its coefficient make a product; multiply each term in the parentheses by the coefficient	use the distributive property to remove parentheses; combine like terms, use inverse operations
Part 11 Inequalities	< > = − + ≤ ≥	inequalities, inequality symbol, greater than, less than, equal to, greater than or equal to, less than or equal to	solve, multiply, divide, change	similar to an equation but has an inequality symbol between each side; greater than, less than, greater than or equal to, less than or equal to	solve like equations but switch the inequality symbol when you divide or multiply both sides by a negative number

2. Expanding algebraic thinking requires further exploration with math concepts in concrete ways. Here are a few suggestions for activities that allow students to explore the concepts in each part of this section of Unit 2.

Part 6: Addition and Subtraction Equations	Separate students into pairs and give each student 20 bingo chips (or other small objects), a piece of paper, and cards with the symbols plus, minus, equal, and *n* variable. On the paper, have each student create an equation that uses both addition and subtraction with the chips and symbol cards (5 chips + 4 chips = 10 chips – 1 chip). The main rule is that the expressions on either side of the equal sign must be equal. Once each student has an equation, have students remove one pile of chips and replace it with the *n* variable card. Then have students switch their equation with a partner. Together, students can help each other solve their equations, discovering how inverse operations work.
Part 7: Multiplication and Division Equations	Expand the above activity by increasing the number of chips students have and exchanging the plus and minus cards for times (use •) and fraction bar cards. Have students repeat the above activity by creating an equation that uses both multiplication and division and has an equal value on either side of the equal sign.
Part 6: Inequalities	The Graphing Inequalities Tools lesson (p. 70) helps students better conceptualize inequalities by seeing them on a number line. Expand this lesson to include more physical practice with inequalities, such as coming up to the board to graph the inequality or using a floor number line that students and on or walk along to show an inequality.

3. Mathematical models guide and remind students of how to apply mathematical concepts. Using models that the students create is a great way to build both student confidence and memory of the concepts. For example, in Strategy 2, students create and solve their own equations. Students can capture this activity in their notebooks by illustrating the equations they created with the bingo chips and showing how they solved them. These then become the models they can refer to when they approach other algebraic equations.

4. One warning to students making their own mathematical models is that they must be accurate representations of the math concept. Incorporating peer evaluation and a final teacher check into this process is a good way to ensure accuracy and provide the opportunity for students to both defend their mathematical reasoning and probe the reasoning of others. A simple checklist can guide this evaluation. If students answer No to any of the questions, they need to adjust their model to meet all criteria.

Evaluation Questions	Yes	No
Is the illustration in your example clear?		
Does your example use both addition and subtraction? -OR- Does your example use both multiplication and division?		
Does the left side of your example equal the right side?		
Does your example replace a value with a variable?		
Does your example show step-by-step how to solve for the variable using inverse operations?		
Does your example help you remember how to solve an equation?		

➢ Bridging Vocabulary

Strategy 1: Identify the component parts and usage of new words to interpret their meanings.

Strategy 2: Use context clues to interpret new words.

Strategy 3: Utilize vocabulary-building resources.

Strategy 4: Build a deeper knowledge of words through math application tasks and collaborative discussions.

Strategy 5: Memorize words through repetitive study such as using flashcards (digital or print) and notes.

1. First, present the shortest form of the word, referred to in this text as the "base word" in the case of academic words and some subject-specific terms. Follow the base form with other commonly used word forms (if available). Examine prefixes and suffixes and their impact on word meaning and part of speech.
2. Read the word as used in the context of the text and discuss possible meanings given context clues and word form.
3. Have students find (electronically or in print) the definition or translation of the base form and, if different, the form used in context and note these definitions in the space provided for future reference and study.
4. Gradually build a deeper knowledge of the word by having students use the word in a sentence frame, guided discussion, and an original sentence within a mathematical context.

Sentence Frame:	*An **inequality** shows that* ____________________.
Guided Discussion:	*When would we use **inequalities** in our lives?*
Original Math Sentence:	____________________

Encourage students to use these words in math applications and collaborative discussions such as the task described in Bridging Problem Solving Strategy 4.

5. The high volume of mathematical terminology requires repeat exposure to the words over time. Word walls, intentionally including the words in questions to students and when eliciting responses from them, and explicit reminders to use the vocabulary in verbal tasks provide built-in reinforcement. However, this is often not enough so it is important that students learn ways to study words independently. Flashcards or websites that offer repetitive vocabulary practice are excellent ways for students to do this. Students may also use their notes, however, they will need to do repetitive activities, similar to flashcard practice, and not simply read and reread their notes.

➤ Bridging Math Application

Strategy 1: ***Prepare for math applications by identifying the problem type and the problem-solving strategies and tools.***

Strategy 2: ***Organize the problem using visual, symbolic, and written representations.***

Strategy 3: ***Overcome barriers to problem solving using math models, language and structural analysis, and resources.***

Strategy 4: ***Demonstrate and defend problem-solving application and mathematical reasoning through reverse problem solving, mental mathematics, visual representations, and peer discussions.***

1. Each part of this lesson lends itself to a variety of math application tasks that allow students to synthesize, apply, or extend their mathematical knowledge and skills. Whichever math application task you choose, be sure to orient students to the problem type and the problem-solving strategies and tools they may utilize. The following is an example of student directions for a math application task that synthesizes the concepts developed in this lesson.

Managing Your Free Time

Part 1:
Your teacher gives you a math project to complete in one week. She/he tells you that you must use half of your week's free time to work on the project. How much time will you spend on the math project? First, define what free time means to you and then use that definition to figure out how much free time you will have for the week. Do not include time that you would usually spend sleeping as free time. You may figure out the answer to the problem in any way you can as long as you can show a classmate (in writing or by drawing) the step-by-step process you used to get your answer.

Part 2:
After finding your answer, explain step-by-step to a partner what you did to get your answer. If you used algebraic equations, work backwards through your steps to see if the answer and the steps you used are accurate. If you did not use algebraic equations, work with your partner to create algebraic equations to work through the problem again. Did you get the same answer using algebra? If not, which one is correct? Go back and re-work the problem with your partner to figure out which method, and therefore which answer, is correct. Finally, use reverse problem solving to check your answer.

Before engaging in problem solving, have students analyze the directions to determine what is being asked of them and use this information to determine which strategies to use to complete the task. The following example provides a glimpse of how this may look.

Directions	**Operation or Action** *Signal Words*	**Strategies and Tools**
Your teacher gives you a math project to complete in one week. She/he tells you that you must use half of your week's free time to work on the project. How much time will you spend on the math project?	add; divide by 2	Figure out total free time for one week. Divide by 2.
First, define what free time means to you.		

2. Students will need to plan and organize how to complete the above task. Because there are so many ways to solve a problem such as the one above, encourage students to create their own method for solving it with the one caveat that whichever method they choose, they must show their methods through writing or illustration. Beginning with a graphic organizer that outlines the steps they need to take and gives them a place to record their answers is a great place to start. This may take a lot of guidance to lead them toward an effective plan. For students who struggle with organizing, provide them a template but gradually release responsibility as they get more exposure to project planning.

Possible Organizational Plans:

1. Figure out amounts of categorized daily activities.
2. Start with Monday and figure out one day at a time.
3. Start with 24 hours a day and subtract each daily activity.
4. Use a schedule to plot daily activities.

3. The above task provides students the opportunity to use their peers for help with problem solving. Along with this, students can refer to the math models they created in the Bridging Problem Solving activity for guidance.

4. The above task requires students to use reverse problem solving with a partner to check their answers. Since the task requires that students show the step-by-step process they used, students can start at the bottom of this sketched process and work their way back to the beginning. Inconsistencies, if they are present, will quickly become apparent. When they are discovered, it is often best to jump to the beginning and work the problem through to the point where the inconsistency was discovered. Once ironed out, reverse the process and work backwards once again.

➢ Assessment & Next Steps

Students should complete the practice activities included in each *Math Sense 2: Focus on Problem Solving* lesson. Evaluate which learning goals were not met and remediate by using other resources, such as those identified in the Bridging Knowledge section. Upon successful completion, continue to the next unit.

Unit 3

SOLVING PROBLEMS WITH ALGEBRA

Skills-Based Questions:

1. How can you prepare to solve a word problem? How can you organize the facts and unknowns of the problem to help you solve the problem? *(Part 1)*
2. How do you solve number puzzles and age problems? *(Part 2)*
3. What is the relationship between rate, time, and distance? How do you solve motion problems? *(Part 3)*
4. How do you solve value problems? *(Part 4)*
5. What is the relationship between rate, time, and amount of work done? How do you solve problems about work? *(Part 5)*
6. How do you rewrite formulas to solve for different variables in the formula? Why might you do this? *(Part 6)*

Math Sense 2: Focus on Problem Solving: Part 1, p. 80; Part 2, p. 82; Part 3, p. 86; Part 4, p. 88; Part 5, p. 90; Part 6, p. 96

	Learning Goals:	GED
Knowledge Goals:	1. Describe how to organize the facts and unknowns of a word problem to help solve the problem. *(Part 1)*	A.1.d
	2. Translate number puzzles and age problems into equations. *(Part 2)*	A.1.g
	3. Describe the relationship between rate, time, and distance and how you solve motion problems. *(Part 3)*	A.1.h
	4. Explain how to solve problems in which counted objects have different values. *(Part 4)*	A.1.j
	5. Describe the relationship between rate, time, and work done and how you solve work problems. *(Part 5)*	
	6. Describe the relationship between different variables in a formula and how you rewrite formulas to solve for different variables in the formula. *(Part 6)*	
Problem-Solving Goals:	1. Organize information from a word problem into a chart and solve. *(Part 1)*	A.1.d
	2. Solve number puzzles and age problems by translating words to symbols. *(Part 2)*	A.1.g
	3. Solve motion problems by finding the rate, time, or distance. *(Part 3)*	A.1.h
	4. Solve value problems given the total number of objects or values. *(Part 4)*	A.1.j
	5. Solve work problems by finding the rate, time, or amount of work done. *(Part 5)*	
	6. Rewrite formulas to isolate the variable you are solving for. *(Part 6)*	
Vocabulary Goals:	1. Define key mathematical terms.	
	2. Determine the meaning of unknown vocabulary using context clues, word forms, and parts of speech.	
	3. Apply new vocabulary to mathematical tasks and discussions.	
Math Application Goals:	1. Apply knowledge of algebraic language to solve real-life math problems. *(Parts 1–6)*	A.1.d
	2. Defend math applications and reasoning to others. *(Parts 1–6)*	A.1.g
		A.1.h
		A.1.j

Sample Instructional Support Strategies

➢ Bridging Knowledge

Strategy 1: ***Develop and connect background knowledge, skills, and conceptual understanding to new knowledge.***

Strategy 2: ***Use guiding questions to make connections beyond the lesson to broader math applications.***

Strategy 3: ***Use problem-solving strategies to develop, monitor, and synthesize conceptual understanding and fluency.*** *(See also Bridging Problem Solving)*

Strategy 4: ***Extend problem-solving skills and mathematical reasoning to broader math applications in life and work.*** *(See Bridging Math Application)*

1. Evaluate students' knowledge of the following mathematical skills. Utilize the chart below to develop student content knowledge as necessary.

Unit 3 **Solving Problems with Algebra**	**Part 1** Translating Words to Equations	**Part 2** Number Puzzles and Age Problems	**Part 3** Solving Motion Problems	**Part 4** Solving Value Problems	**Part 5** Solving Work Problems	**Part 6** Rewriting Formulas
Core Skills in Mathematics	Unit 4, Lesson 3: Solving Equations (p. 64); Solving Problems Using Expressions, Equations, and Inequalities (p. 76)					Graphing Equations and Inequalities, p. 72
Scoreboost Mathematics: Algebraic Reasoning	Write and Solve Equations, p. 14					Solve Inequalities, p. 18
Pre-HSE Workbook: Math 2	Solving Equations, p. 18		Solving Multistep Equations, p. 20			Solving Inequalities, p. 28

2. The unit preview for each unit in the *Math Sense 2: Focus on Problem Solving* provides a list of context examples in which to apply the target math concepts as well as questions that connect to students' prior knowledge and experience (for Unit 3, p. 79). These questions also provide a bridge to broader math applications in life and work.

3. Each unit also provides focus lessons on problem solving, using tools, and test taking. Use these lessons to bridge student knowledge in these areas.

Unit 3	**Content**	**Page**
Problem Solver	Solve Equations with Fractions and Decimals Using Inequalities to Solve Word Problems	84 92
Tools	NA	
Test Taker	Using Formulas	94

➢ Bridging Problem Solving

Strategy 1: ***Preview the problem to determine problem-solving strategies and tools and predict general solutions.***

Strategy 2: ***Develop conceptual understanding of mathematical problems using visual representations, think-alouds, and collaboration.***

Strategy 3: ***Overcome barriers to problem solving using math models, language and structural analysis, and resources.***

Strategy 4: ***Demonstrate and defend problem solving and mathematical reasoning through reverse problem solving, mental mathematics, visual representations, and peer discussions.***

1. For Parts 1–6 of Unit 3, orient students to features of each mathematical concept such as the symbols, language, and structure. Students should identify the math language and symbols to determine what is being asked of them and use this information to determine which strategies to use to complete the task. Although it is not necessary for students to fill out a graphic organizer for each math problem they attempt, completing the following graphic organizer is helpful for annotating and reviewing math concepts and choosing appropriate strategies to complete tasks. The following is an example of how a student (with guidance) might fill out this graphic organizer:

	Symbols	**Language**	**Operation or Action**	**Structure**	**Strategies/Tools**
Part 1 Translate Words to Equations	× ÷ + − ( ) $\frac{\blacksquare}{\blacksquare}$ =	fact, unknown, chart, multiplication, division, addition, subtraction, equation, variable, chart	create, organize, write, solve	facts are known information; use variables for unknowns	use a chart to organize information: facts and unknowns; use words to find relationships
Part 2 Number Puzzles and Age Problems	× ÷ + −	product, quotient, difference, sum	multiply, divide, subtract, add	n = unknown; is means equals; clues tell operations	use a chart to organize; identify the unknown as n; write out the clues as operations; solve
Part 3 Motion Problems	+ − × ÷ ( ) $\frac{\blacksquare}{\blacksquare}$ =	rate, time distance, formula, how far, how long, how fast, what speed	multiply, divide, solve	$r \times t = d$	use the formula and a chart or picture to organize problem
Part 4 Value Problems	+ − ( ) ÷ = ×	value, number, worth, amount, total, how many, how much	add, subtract, multiply, divide, combine	use variable x for the unknown; add facts to unknown ($2x$); use total number or value on other side of equal sign	use a chart to organize the numbers and values; use a variable for unknown; subtract variable from total amount to find second unknown
Part 5 Work Problems	+ − × ÷ ( ) $\frac{\blacksquare}{\blacksquare}$ = n	work, rate, time, fraction, denominator, numerator, reciprocal, how long,	multiply, divide, solve	work formula is like motion formula: $r \times t$ = work	use the formula and a chart or picture to organize problem
Part 6 Rewrite Formulas	+ − × ÷ ( ) $\frac{\blacksquare}{\blacksquare}$ = n	equivalent formulas, formula, variables, inverse operations; isolate	rewrite, isolate, solve	each variable may be isolated by using inverse operations to rewrite formulas	use inverse operations to rewrite formulas to isolate the variable you are solving for

2. Expanding algebraic thinking requires further exploration with math concepts in concrete ways. Here is an activity that allows students to explore the concepts in Unit 3.

Preparing for Exploration:	Before beginning the activity, prepare the class to isolate the facts and unknowns from a word problem into a graphic organizer such as a chart. Give students a paper with three to five word problems on it and two different colored highlighters (or have them circle or underline). Read the first word problem together. Students highlight with one color (or circle) the facts, selecting only the important words and numbers. Then students highlight with the other color (or underline) the unknown in the question (again, just the relevant information). Help students construct a chart* with this information. For problems with only one unit, each individual item/person/event will become a column heading and the respective facts and unknown (x) will be placed below their respective heading. If the problem deals with more than one unit, the units will occupy the column headings and each item/person/event will occupy the row headings. Each fact and unknown (x) will occupy the intersection of its unit and description. Continue using the same process as you move through the problems, gradually releasing responsibility to the students for identifying facts and unknowns and creating organizational charts. *For ELLs, chart work can be quite challenging. It may be necessary to explicitly teach how to make and read a chart.
Station Work:	Create five stations for students to practice each of type of word problem presented in *Math Sense 2: Focus on Problem Solving* Unit 3: number puzzles, age problems, motion problems, value problems, and work problems. Divide students into groups that will rotate through the stations. At each station, provide a model for that type of word problem, which includes a completed chart, the written equation, and the solution to the equation. Students then work together to solve two additional problems at the station. Each group will keep their work with them as they travel to the next station.
Evaluation	After students complete each station, have each group meet with another group and share the problems they attempted to solve. Have them compare their charts, equations, and solutions. Have them discuss the accuracy or inaccuracy of the work and negotiate a final answer.
Lesson Introduction:	Collect the final answers from the groups and use them to introduce each lessons in Unit 3. Beginning with the student work, explore the mathematical reasoning behind the student work and compare it to the lesson instructions.

3. Understanding word problems is directly tied to making sense of the words within them. This can be especially difficult for ELLs. As students translate word problems into equations, have them refer to a list of signal words for math operations such as those on p. 159 of this book. Having this reference guide available at each station during the above activity would also be helpful.

4. When groups of students meet to compare their problem-solving strategies and solutions such as in the above activity, they are accountable for applying their previously practiced algebraic skills, negotiating problem-solving processes and solutions, and using mathematical language. Establishing norms for this kind of group work is important. Here is a simple list of norms you may want to remind students of before a group activity like this one.

1. Each person **participates** equally.	**Stop** if you are doing too much and **let others** practice. **Start working** if you are watching too much.
2. Each person's ideas will be **listened to.**	**Stop** if you are talking too much and **listen** to others. **Talk** if you are not talking enough and **share** your ideas.
3. Each person will use the **mathematical language** practiced in class.	**Look** at your language paper if you need help. Politely **help** your teammates use the correct language.
4. Each person has the right to **politely disagree.**	**Look** at your language paper about polite disagreement. **Explain** why you disagree.

➢ Bridging Vocabulary

Strategy 1: ***Identify the component parts and usage of new words to interpret their meanings.***

Strategy 2: ***Use context clues to interpret new words.***

Strategy 3: ***Utilize vocabulary-building resources.***

Strategy 4: ***Build a deeper knowledge of words through math application tasks and collaborative discussions.***

Strategy 5: ***Memorize words through repetitive study such as using flashcards (digital or print) and notes.***

1. First, present the shortest form of the word, referred to in this text as the "base word" in the case of academic words and some subject-specific terms. Follow the base form with other commonly used word forms (if available). Examine prefixes and suffixes and their impact on word meaning and part of speech.
2. Read the word as used in the context of the text and discuss possible meanings given context clues and word form.
3. Have students find (electronically or in print) the definition or translation of the base form and, if different, the form used in context and note these definitions in the space provided for future reference and study.
4. Gradually build deeper knowledge of the word by having students use the word in a sentence frame, guided discussion, and an original sentence within a mathematical context.

Sentence Frame:	***Rate*** *helps us calculate* ____________________.
Guided Discussion:	*In which situations in life do we use* ***rate****?*
Original Math Sentence:	____

Encourage students to use these words in math applications and collaborative discussions such as the task described in Bridging Problem Solving Strategy 4.

5. The high volume of mathematical terminology requires repeat exposure to the words over time. Word walls, intentionally including the words in questions to students and when eliciting responses from them, and explicit reminders to use the vocabulary in verbal tasks provide built-in reinforcement. However, this is often not enough so it is important that students learn ways to study words independently. Flashcards or websites that offer repetitive vocabulary practice are excellent ways for students to do this. Students may also use their notes, however, they will need to do repetitive activities, similar to flashcard practice, and not simply read and reread their notes.

➢ Bridging Math Application

Strategy 1: ***Prepare for math applications by identifying the problem type and the problem-solving strategies and tools.***

Strategy 2: ***Organize the problem using visual, symbolic, and written representations.***

Strategy 3: ***Overcome barriers to problem solving using math models, language and structural analysis, and resources.***

Strategy 4: ***Demonstrate and defend problem-solving application and mathematical reasoning through reverse problem solving, mental mathematics, visual representations, and peer discussions.***

1. Each part of this lesson lends itself to a variety of math application tasks that allow students to synthesize, apply, or extend their mathematical knowledge and skills. Whichever math application task you choose, be sure to orient students to the problem type and the problem-solving strategies and tools they may utilize. The following is an example of student directions for a math application task that synthesizes the concepts developed in this lesson.

Choosing a Job

Last week you interviewed for two jobs*. Today, you found out that both companies want to hire you. Which job will you take? Compare the jobs for pay, benefits, hours, and distance from your home. Choose the job that provides the most financial benefit to you. Make sure to consider in your choice the cost (in time and money) of travel to and from the position. As you work through the problem, write out each step you take and be prepared to explain the mathematical reasoning you used to solve the problem.

Job 1:

40 hours/week

$15.65/hour

5 paid sick days/year

1 week paid vacation/year

$450/month paid toward a $700/month health plan

35 miles from home (1 way)

$______ /gallon price of gas

Job 2:

35 hours/week

$18.25/hour

No sick pay

3 paid vacation days/year

$250/month paid toward a $600/month health plan

7 miles from home (1 way)

$______ /gallon price of gas

* For an authentic application of this problem, have students research local jobs they are qualified for and use the actual information from those jobs to solve this problem.

Before engaging in problem solving, have students analyze the directions to determine what is being asked of them and use this information to determine which strategies to use to complete the task. The following example provides a glimpse of how this may look.

Directions	Operation or Action *Signal Words*	Strategies and Tools
Compare the jobs for pay, benefits, hours, and distance from your home.	multiply, add, subtract, divide	Calculate each fact of each job separately; line up to compare; make sure all time units are equal when comparing
Choose the job that provides the most financial benefit to you.		

2. Students will need to organize how to complete the above task. Because there are so many ways to solve a problem such as the one above, encourage students to create their own method for solving it with the one caveat that whichever method they choose, they must show their methods through writing or illustration. A graphic organizer that outlines the steps they need to take and gives them a place to record their answers is a great place to start. This may take a lot of guidance to lead them toward an effective plan. For students who struggle with organizing, provide them a template but gradually release responsibility as they get more exposure to project planning.

Job 1*:

Hours (including drive time)	**Pay**	**Sick/Vacation Paid Time Off**	**Health Insurance**	**Cost of Gas**
Total:	Total:	Total:	Total:	Total:
Total Pay (including benefits):				
Total Cost (including health insurance & gas):				

***Repeat for Job 2**

3. Students can collaborate with others to organize and solve the problem. They can continue to use the algebraic models they created in the previous unit to guide them in solving this problem.

4. Once students choose the job they believe provides the most financial benefit, have them meet with a partner and defend their choice by explaining the mathematical process and reasoning that lead up to their choice. When students disagree in a particular area, they need to respectfully negotiate a resolution to their disagreement. This resolution should be solidly grounded in mathematical evidence.

➢ Assessment & Next Steps

Students should complete the practice activities included in each *Math Sense 2: Focus on Problem Solving* lesson. Evaluate which learning goals were not met and remediate by using other resources, such as those identified in the Bridging Knowledge section. Upon successful completion, continue to the next unit.

Unit 4

GEOMETRY BASICS: Section 1

Skills-Based Questions:

1. What terminology do we use to discuss basic geometry problems? *(Part 1)*
2. What are the different types of angles and how do we find their measurements? *(Part 2)*
3. What types of quadrilaterals are there and how do we find the measurements of their angles? *(Part 3)*
4. What are the different types of triangles and how do we find the measurements of their angles? *(Part 4)*
5. What is the Pythagorean Theorem and why and how do we use it? *(Part 5)*
6. How can knowing information about one triangle or rectangle help you find information about a similar one? *(Part 6)*

Math Sense 2: Focus on Problem Solving: Part 1, p. 102; Part 2, p. 106; Part 3, p. 108; Part 4, p. 110; Part 5, p. 112; Part 6, p. 114

	Learning Goals:	**GED**
Knowledge Goals:	1. Define basic geometry terms needed to solve geometry problems. *(Part 1)*	Prep Q.4.a
	2. Describe the different types of angles and how their relationships help you find their measurements. *(Part 2)*	Prep Q.4.e
	3. Describe types of quadrilaterals and how the relationship between the sides and angles of a quadrilateral help you find an unknown measurement. *(Part 3)*	
	4. Describe types of triangles and how the relationship between the sides and angles of a triangle help you find an unknown measurement. *(Part 4)*	
	5. Explain the Pythagorean Theorem and how to use it to find an unknown length of a right triangle. *(Part 5)*	
	6. Describe how proportions can help you find the measurements of similar triangles and rectangles. *(Part 6)*	
Problem-Solving Goals:	1. Interpret basic geometry terms to understand and solve geometry problems. *(Part 1)*	Prep Q.4.a
	2. Use the known properties of angles to find unknown angle measurements. *(Part 2)*	Prep Q.4.e
	3. Find measurements of the sides and angles of quadrilaterals using knowledge of quadrilateral properties. *(Part 3)*	
	4. Find measurements of the sides and angles of triangles using knowledge of triangle properties. *(Part 4)*	
	5. Use the Pythagorean Theorem to find the unknown length of a right triangle. *(Part 5)*	
	6. Use proportions to find measurements of similar triangles and rectangles. *(Part 6)*	
Vocabulary Goals:	1. Define key mathematical terms.	
	2. Determine the meaning of unknown vocabulary using context clues, word forms, and parts of speech.	
	3. Apply new vocabulary to mathematical tasks and discussions.	
Math Application Goals:	1. Apply knowledge of triangles and quadrilaterals to solve real-life math problems. *(Parts 1–6)*	Prep Q.4.a
	2. Defend math applications and reasoning to others. *(Parts 1–6)*	Prep Q.4.e

Sample Instructional Support Strategies

➢ Bridging Knowledge

Strategy 1: ***Develop and connect background knowledge, skills, and conceptual understanding to new knowledge.***

Strategy 2: ***Use guiding questions to make connections beyond the lesson to broader math applications.***

Strategy 3: ***Use problem-solving strategies to develop, monitor, and synthesize conceptual understanding and fluency.*** *(See also Bridging Problem Solving)*

Strategy 4: ***Extend problem-solving skills and mathematical reasoning to broader math applications in life and work.*** *(See Bridging Math Application)*

1. Evaluate students' knowledge of the following mathematical skills. Utilize the chart below to develop student content knowledge as necessary.

Unit 4 **Geometry Basics, Section 1**	**Part 1** Points, Lines, and Angles	**Part 2** Working with Angles	**Part 3** Quadrilaterals	**Part 4** Triangles	**Part 5** The Pythagorean Theorem	**Part 6** Similar Geometric Figures
Core Skills in Mathematics	Unit 5, Lesson 1: Knowing Shapes and Their Attributes (p. 84)	Unit 5, Lesson 1: Knowing Shapes and Their Attributes (p. 84); Lesson 3: Solving Angle Measure Problems (p. 94)				
Scoreboost Mathematics: Measurement and Geometry					Use the Pythagorean Theorem with Right Triangles (p. 36)	
Pre-HSE Workbook: Math 1					Using the Pythagorean Theorem (p. 36)	
Pre-HSE Workbook: Math 2	Solving Equations (p. 18); Solving Multi-Step Equations (p. 20)					

2. The unit preview for each unit in the *Math Sense 2: Focus on Problem Solving* provides a list of context examples in which to apply the target math concepts as well as questions that connect to students' prior knowledge and experience (for Unit 4, p. 101). These questions also provide a bridge to broader math applications in life and work.

3. Each unit also provides specialized lessons that focus on problem solving, using tools, and test taking techniques using the math skills taught in the lesson. Utilize these lessons to build student knowledge in these areas.

Unit 4, Section 1	Content	Page
Problem Solver	Finding Patterns in Algebra and Geometry	116
Tools	Using Protractors	104
Test Taker	NA	

➢ Bridging Problem Solving

Strategy 1: ***Preview the problem to determine problem-solving strategies and tools and predict general solutions.***

Strategy 2: ***Develop conceptual understanding of mathematical problems using visual representations, think-alouds, and collaboration.***

Strategy 3: ***Overcome barriers to problem solving using math models, language and structural analysis, and resources.***

Strategy 4: ***Demonstrate and defend problem solving and mathematical reasoning through reverse problem solving, mental mathematics, visual representations, and peer discussions.***

1. For Parts 1–6 of Unit 4, orient students to features of each mathematical concept such as the symbols, language, and structure. Students should identify the math language and symbols to determine what is being asked of them and use this information to determine which strategies to use to complete the task. Although it is not necessary for students to fill out a graphic organizer for each math problem they attempt, completing the following graphic organizer is helpful for annotating and reviewing math concepts and choosing appropriate strategies to complete tasks. The following is an example of how a student (with guidance) might fill out this graphic organizer:

	Symbols	Language	Operation or Action	Structure	Strategies/Tools
Part 1 Points, Lines, and Angles	° •	point, line, ray, line segment, vertical, horizontal, intersect, rotation, parallel, perpendicular, vertex (vertices), right, acute, straight, obtuse, degrees	identify, name, measure	a line is named by single italic letter; a line segment is named by the capital letters of its start and end points; a ray is named by its start point and the letter it passes through	identify kinds of lines and how they are labeled; identify four kinds of angles, their rotation, and what they look like
Part 2 Working with Angles	°	right angle, sum, complementary, supplementary, straight angle, degrees	divide, add, subtract	complimentary: sum of angles = 90°; supplementary: sum of angles = 180°; intersecting lines: sum of adjacent and vertical angles = 360°	identify kinds of angles; subtract 1 angle from known total (based on angle properties) to find the other angle
Part 3 Quadrilaterals	° •	polygon, quadrilateral, opposite, adjacent, diagonal, square, rectangle, rhombus, parallelogram, trapezoid, degrees	identify, name, measure,	quadrilaterals are named by the four letters of the vertices; sum of all angles = 360°	identify kinds of quadrilaterals; use their properties to find measurement of angles
Part 4 Triangles		triangle, equilateral, isosceles, right, scalene	identify, measure, solve	sum of angles in a triangle = 180°; equal sides are opposite equal angles; longest side is opposite largest angle	identify kinds of triangles; use their properties to solve for unknown angles
Part 5 Pythagorean Theorem	a^2 $\sqrt{\ }$	right angle, legs, hypotenuse	substitute, evaluate, simplify, solve, find	$a^2 + b^2 = c^2$; sum of the squares of the legs of a right triangle = the square of the hypotenuse	use the Pythagorean Theorem to find the length of the unknown side
Part 6 Similar Geometric Figures	≈	similar figures, ratio, proportion, length, width	set up, cross multiply, solve	set up a proportion with corresponding sides; cross multiply to solve	use proportions to solve for unknown of a similar triangle or quadrilateral

2. Expanding algebraic thinking requires further exploration with math concepts in concrete ways. Here is an activity that allows students to explore the concepts of this section of Unit 4 through an interactive scavenger hunt.

Preparation:

Students will need the following materials for the scavenger hunt:

1. A notebook and pencil to draw shapes and record measurements
2. A measuring tape
3. A protractor
4. Scavenger hunt list*

Before this activity, teach students the names of the shapes on the scavenger hunt list and how to use the measuring tape and protractor. Explain that for this activity, they will measure width and length to the nearest ¼ inch and angles to the nearest ½ degree.

*Make sure that there is something in the classroom that represents each of the shapes on the list.

Scavenger Hunt List:

- line
- angle (open/not triangle)
- square
- rectangle
- parallelogram
- right triangle
- triangle with 3 acute angles
- triangle with 1 obtuse angle

Student Directions:

1. Find each item on the list in the room (or building).
2. Draw the item in your notebook.
3. Measure each side of the item and record the measurements in the correct places on the drawing.
4. Measure each angle and record the measurements on the drawing next to the correct angle.

Full-Class Discussion:

Draw the shapes on the board. Have students share the names of the items they found for each shape. Then have students share the measurements of the sides and angles of those items. Record all information visually on the board. Lead the class in exploring the relationship of the sides and angles of the different triangles and quadrilaterals. What can they tell you about the total of the angles of a triangle? A quadrilateral? What can they tell you about the sides opposite large or small angles in a triangle? Discover through this activity the properties of the different kinds of triangles and quadrilaterals. Use this to lead into each corresponding lesson in Unit 4.

3. As students participate in the discussion above, have them take notes next to their drawings in their notebooks. Students should label the names of the shapes and the properties they discover through the discussion. This notebook information then becomes a reference model as they continue through the unit.

4. The full-class discussion above gives students the opportunity to develop their conceptual understanding through making observations and answering targeted questions. Check in with each student during this discussion to ensure that they are engaged in making the connections being discussed and taking adequate notes in their notebooks. Ask questions that elicit the kinds of discoveries that will help students with the upcoming lessons.

➤ Bridging Vocabulary

Strategy 1: Identify the component parts and usage of new words to interpret their meanings.

Strategy 2: Use context clues to interpret new words.

Strategy 3: Utilize vocabulary-building resources.

Strategy 4: Build a deeper knowledge of words through math application tasks and collaborative discussions.

Strategy 5: Memorize words through repetitive study such as using flashcards (digital or print) and notes.

1. First, present the shortest form of the word, referred to in this text as the "base word" in the case of academic words and some subject-specific terms. Follow the base form with other commonly used word forms (if available). Examine prefixes and suffixes and their impact on word meaning and part of speech.
2. Read the word as used in the context of the text and discuss possible meanings given context clues and word form.

3. Have students find (electronically or in print) the definition or translation of the base form and, if different, the form used in context and note these definitions in the space provided for future reference and study.
4. Gradually build a deeper knowledge of the word by having students use the word in a sentence frame, guided discussion, and an original sentence within a mathematical context.

Sentence Frame:	*An* ***angle*** *is formed when* ____________________.
Guided Discussion:	*When in life do you need to know the measurement of an* ***angle****? Why?*
Original Math Sentence:	____

Encourage students to use these words in math applications and collaborative discussions such as the task described in Bridging Problem Solving Strategy 4.

5. The high volume of mathematical terminology requires repeat exposure to the words over time. Word walls, intentionally including the words in questions to students and when eliciting responses from them, and explicit reminders to use the vocabulary in verbal tasks provide built-in reinforcement. However, this is often not enough so it is important that students learn ways to study words independently. Flashcards or websites that offer repetitive vocabulary practice are excellent ways for students to do this. Students may also use their notes, however, they will need to do repetitive activities, similar to flashcard practice, and not simply read and reread their notes.

➢ Bridging Math Application

Strategy 1: Prepare for math applications by identifying the problem type and the problem-solving strategies and tools.

Strategy 2: Organize the problem using visual, symbolic, and written representations.

Strategy 3: Overcome barriers to problem solving using math models, language and structural analysis, and resources.

Strategy 4: Demonstrate and defend problem-solving application and mathematical reasoning through reverse problem solving, mental mathematics, visual representations, and peer discussions.

1. Each part of this lesson lends itself to a variety of math application tasks that allow students to synthesize, apply, or extend their mathematical knowledge and skills. Whichever math application task you choose, be sure to orient students to the problem type and the problem-solving strategies and tools they may utilize. The following is an example of student directions for a math application task that synthesizes the concepts developed in this lesson.

Design a Mural

Part 1:
The director of your school wants to add more art to the school's interior. She has asked each class in the school to submit a design for a mural that will be painted in the school lobby. Students will vote on the designs and the one with the most votes will be used for the lobby mural. Your teacher wants the design for your class to reflect the geometric shapes you have been learning. In a small group, create a drawing of a design you would like to submit for the contest. Your design must fit on an $8\frac{1}{2} \times 11$ sheet of paper. It must contain only triangles and quadrilaterals. This means that if there is space between objects, this space must also be in the form of triangles or quadrilaterals. Apply color to the shapes with colored pencils, markers, crayons, or paint. Finally, label each shape with the actual measurements of its sides and angles so its proportions can be used to create the large mural.

Part 2:
Once all of the designs are completed, have the class vote on the three designs they like the most. They should put their choices in order of preference. Once the top design is chosen, together (as a class) determine the maximum space the mural will occupy. Develop a proportion using the outside measurements of the design that will fit within that maximum space. Use this proportion to set the scale for the project. Compute the lengths and widths of the shapes in the design that will be needed to make the mural.

Before engaging in problem solving, have students analyze the directions to determine what is being asked of them and use this information to determine which strategies to use to complete the task. The following example provides a glimpse of how this may look.

Directions	Operation or Action *Signal Words*	Strategies and Tools
In a small group, create a drawing of a design you would like to submit for the contest. Your design must fit on an 8½ × 11 sheet of paper. It must contain only triangles and quadrilaterals.	create, draw	Use properties of triangles and squares to plan the design.
Apply color to the shapes with colored pencils, markers, crayons, or paint.		

2. Students will need to organize how to complete the above task. This particular activity does not involve problem solving, but rather design. As such, a step-by-step checklist should help students meet the requirements of the project.

    1. Use an 8½ × 11 sheet of paper.
    2. Draw a design with all triangles and quadrilaterals.
    3. Even the spaces (if used) between shapes must be triangles and quadrilaterals.
    4. Color the shapes.
    5. Measure the sides and angles of the different shapes in the design.
    6. Label the measurements on the design.

3. Students can refer to the models of shapes they have in their notebooks, using the properties of the different triangles and quadrilaterals to help them create their designs. They can also refer to the lessons in Unit 4 of *Math Sense 2: Focus on Problem Solving*.

4. Voting on the designs as a full class gives students the opportunity to evaluate other groups' designs in order to choose the design that meets the requirements of the assignment but is also the most appealing. After the design is chosen, students need to calculate the actual dimensions of the mural. Once small groups have completed this process, two groups can come together to compare their calculations and evaluate their accuracy. Students will need to refer to the properties of the shapes to iron out differences in calculations.

➢ Assessment & Next Steps

Students should complete the practice activities included in each *Math Sense 2: Focus on Problem Solving* lesson. Evaluate which learning goals were not met and remediate by using other resources, such as those identified in the Bridging Knowledge section. Upon successful completion, continue to the next section of the unit.

GEOMETRY BASICS: Section 2

Skills-Based Questions:

1. How does knowing the properties of triangles and quadrilaterals help us find the perimeter of polygons? *(Part 7)*
2. What is area and how do you calculate the area of quadrilaterals? *(Part 8)*
3. How do you calculate the area of triangles and trapezoids? *(Part 9)*
4. What is *pi*? How do you use it to find the circumference and area of circles? *(Part 10)*
5. What is volume? How do you find the volume of prisms and cylinders? *(Part 11)*
6. How do you calculate the volume of pyramids, cones, and spheres? *(Part 12)*
7. What is surface area and how do you calculate it for 3-dimensional figures? *(Part 13)*

Math Sense 2: Focus on Problem Solving: Part 7, p. 120; Part 8, p. 122; Part 9, p. 124; Part 10, p. 126; Part 11, p. 130; Part 12, p. 132; Part 13, p. 134

	Learning Goals:	**GED**
Knowledge Goals:	1. Describe how knowing the properties of triangles and quadrilaterals helps you find perimeter. *(Part 7)* 2. Describe how to calculate the area of quadrilaterals. *(Part 8)* 3. Explain how to calculate the area of triangles and trapezoids by using knowledge of the properties of parallelograms and triangles. *(Part 9)* 4. Explain what pi is and how it is used to find the circumference and area of circles. *(Part 10)* 5. Describe how to calculate the volume of prisms and cylinders. *(Part 11)* 6. Describe how to calculate the volume of pyramids, cones, and spheres. *(Part 12)* 7. Describe surface area and how to calculate it for 3D shapes. *(Part 13)*	Q.4.a Q.4.b Q.4.c Q.5.a Q.5.b Q.5.a Q.5.b Q.5.c Q.5.d Q.5.e
Problem-Solving Goals:	1. Find the perimeter of polygons using knowledge of triangle and quadrilateral properties. *(Part 7)* 2. Calculate the area of quadrilaterals. *(Part 8)* 3. Calculate the area of triangles and trapezoids. *(Part 9)* 4. Calculate the circumference and area of circles using formulas with pi. *(Part 10)* 5. Calculate the volume of prisms and cylinders using formulas. *(Part 11)* 6. Calculate the volume of pyramids, cones, and spheres using formulas. *(Part 12)* 7. Calculate the surface area of common 3D shapes. *(Part 13)*	Q.4.a Q.4.b Q.4.c Q.5.a Q.5.b Q.5.a Q.5.b Q.5.c Q.5.d Q.5.e
Vocabulary Goals:	1. Define key mathematical terms. 2. Determine the meaning of unknown vocabulary using context clues, word forms, and parts of speech. 3. Apply new vocabulary to mathematical tasks and discussions.	
Math Application Goals:	1. Apply knowledge of perimeter, area, and volume to solve real-life math problems. *(Parts 7–13)* 2. Defend math applications and reasoning to others. *(Parts 7–13)*	Q.4.a Q.4.c Q.5.a Q.5.b Q.5.c Q.5.d Q.5.e

Sample Instructional Support Strategies

➢ Bridging Knowledge

Strategy 1: ***Develop and connect background knowledge, skills, and conceptual understanding to new knowledge.***

Strategy 2: ***Use guiding questions to make connections beyond the lesson to broader math applications.***

Strategy 3: ***Use problem-solving strategies to develop, monitor, and synthesize conceptual understanding and fluency.*** *(See also Bridging Problem Solving)*

Strategy 4: ***Extend problem-solving skills and mathematical reasoning to broader math applications in life and work.*** *(See Bridging Math Application)*

1. Evaluate students' knowledge of the following mathematical skills. Utilize the chart below to develop student content knowledge as necessary.

Unit 4 **Geometry Basics, Section 2**	**Part 7** Perimeter	**Part 8** Area of Squares, Rectangles, and Parallelograms	**Part 9** Area of Triangles and Trapezoids	**Part 10** Circumference and Area of Circles	**Part 11** Volume of Prisms and Cylinders	**Part 12** Volume of Pyramids, Cones, and Spheres	**Part 13** Surface Area
Core Skills in Mathematics	Unit 5, Lesson 4: Solving Perimeter and Area Problems (p. 99)				Unit 5, Lesson 5: Solving Surface Area and Volume Problems (p. 103)		
Scoreboost Mathematics: Measurement and Geometry	Find Circumference, Perimeter, and Area (p. 14); Fill-in-the-Blank Perimeter, Circumference, or Area Questions (p. 16); Find a Dimension When Given Perimeter, Circumference, or Area (p. 18)				Using the Formula Sheet to Find Volume (p. 24)		Using the Formula Sheet to Find Surface Area (p. 26)
Pre-HSE Workbook: Math 1	Calculating Perimeter and Circumference (p. 28)	Calculating Area (p. 30)		Calculating Perimeter and Circumference (p. 28)	Calculating Volume (p. 34)		Calculating Surface Area (p. 32)

2. The unit preview for each unit in the *Math Sense 2: Focus on Problem Solving* provides a list of context examples in which to apply the target math concepts as well as questions that connect to students' prior knowledge and experience (for Unit 4, p. 101). These questions also provide a bridge to broader math applications in life and work.

3. Each unit also provides specialized lessons that focus on problem solving, using tools, and test taking techniques using the math skills taught in the lesson. Utilize these lessons to build student knowledge in these areas.

Unit 4, Section 2	**Content (GED Target)**	**Page**
Problem Solver	Area of Complex Figures (Q.4.d)	128
	Choosing Area, Perimeter, Volume, or Surface Area	136
Tools	Using Algebra in Geometry Problems	138
Test Taker	NA	

➢ Bridging Problem Solving

Strategy 1: ***Preview the problem to determine problem-solving strategies and tools and predict general solutions.***

Strategy 2: ***Develop conceptual understanding of mathematical problems using visual representations, think-alouds, and collaboration.***

Strategy 3: ***Overcome barriers to problem solving using math models, language and structural analysis, and resources.***

Strategy 4: ***Demonstrate and defend problem solving and mathematical reasoning through reverse problem solving, mental mathematics, visual representations, and peer discussions.***

1. For Parts 7–13 of Unit 4, orient students to features of each mathematical concept such as the symbols, language, and structure. Students should identify the math language and symbols to determine what is being asked of them and use this information to determine which strategies to use to complete the task. Although it is not necessary for students to fill out a graphic organizer for each math problem they attempt, completing the following graphic organizer is helpful for annotating and reviewing math concepts and choosing appropriate strategies to complete tasks. The following is an example of how a student (with guidance) might fill out this graphic organizer:

	Symbols	Language	Operation or Action	Structure	Strategies/Tools
Part 7 Perimeter	+	perimeter, sides, length, add, distance	add, calculate	add all sides together; in parallelograms opposite sides are equal; in isosceles triangles 2 sides are equal; etc.	use properties of triangles and quadrilaterals to figure out lengths of unknown sides
Part 8 Area of Quadrilaterals	× cm^2 sq	square, rectangle, parallelogram, area, surface, multiply, two-dimensional figures, base, height, squared	multiply, square	height is perpendicular (right angle) to the base (horizontal); use height × base to find area	multiply base by height to find area (like counting tiles that cover the surface)
Part 9 Area of Triangles and Trapezoids	× ÷ $\frac{■}{■}$ in^2 sq	triangle, trapezoid, parallelogram, area, base, height, square	multiply, divide, find, solve	A of triangle = ½ bh A of trapezoid = ½ $h(b_1 + b_2)$	use the formulas for area of a triangle and trapezoid to find their areas; substitute values to solve
Part 10 Circumference and Area of Circles	πr^2 ≈	circle, circumference, diameter, radius, pi, formula	find, multiply	circumference = πd, with $\pi \approx 3.14$ $Area = \pi r^2$	use the formulas to find circumference and area of circles; substitute values to solve
Part 11 Volume of Prisms and Cylinders	πr^2 ≈ • cm^3	three-dimensional, cubed, length, width, height, depth, prism, vertex, edge, face, volume, cylinder	multiply, divide, find, solve	volume of prism: $V = Bh$ (B = area of one base; h = height) volume of cylinder: $V = \pi r^2 h$ (r = radius, h = height)	use the formulas to find the volume of prisms and cylinders; substitute values to solve
Part 12 Volume of Pyramids, Cones, and Spheres	⅓ πr^2 ≈ • cm^3	three-dimensional, cubed, length, pyramid, cone, sphere	solve, substitute, find	use volume formulas from the GED math formula sheet	use the formulas to find the volume of pyramids, cones, and spheres; substitute values to solve
Part 13 Surface Area	+ πr^2 ½ in^2	surface area, three-dimensional, faces, squared	multiply, add, find, solve	add areas of the faces of a figure to get surface area; use surface area formulas from the GED math formula sheet	use the formulas to find the surface area of three-dimensional figures; substitute values to solve

2. Expanding algebraic thinking requires further exploration with math concepts in concrete ways. Here is an activity that allows students to explore the concepts of this section of Unit 4 through activity stations.

Preparing for Exploration:

Separate students into groups to explore one of the following geometry topics and share their findings with their classmates. Use the four corners of the room to set up an exhibit for each of the four geometry topics. These exhibits should show hands-on examples of the figures to be explored and their names. Have students visit each corner quickly and then choose a corner that they want to explore further. Encourage students to create equal numbered groups at each corner.

Geometry Areas:

1. Perimeter and area of triangles and quadrilaterals
2. Circumference and area of circles
3. Volume of prisms and cylinders
4. Volume of pyramids, cones, and spheres

Research:

Once a group is established for each of the geometry topics, the members of the group will do additional research about the topic. Students can use text books and websites to find their information. Students will create a poster with the following information regarding their topic of geometry:

1. A clear illustration of each figure at the station
    a. Labeled with its name
    b. Labeled with its important parts (e.g. base, height, diameter, radius, etc.)
2. A written description of the properties of the figure (e.g. parallel sides, equal sides, right angle, etc.)
3. An image of each figure as found in real life
4. Formulas for how to calculate measurements that pertain to that topic (e.g. volume, perimeter, etc.)

Sharing:

Create a Gallery Walk with the student posters. Assign one person per group to stand by its poster and explain to passersby for a 5- to 7-minute stint. Rotate each member of each group through this position until everyone has had a chance to explain his or her group's poster and view the other posters.

3. The above activity allows students to use resources, including their group members, to develop an understanding of their topic. The posters they create can then be used as models to help them delve deeper into each geometry topic.

4. The sharing portion of the above activity gives each student the opportunity to explain his/her geometry topic to other students. This is a great way to both hold students accountable for understanding their topic and for explaining their understanding. Provide a checklist for students when it is their turn to explain their poster.

1. Describe each of the figures on your poster.
    a. Tell the name of each figure.
    b. Tell the names of its parts.
2. Explain the properties of each figure on your poster.
3. Explain where you see each figure in real life.
4. Tell the formulas you can use to calculate measurements of the figures.

➢ Bridging Vocabulary

Strategy 1: Identify the component parts and usage of new words to interpret their meanings.

Strategy 2: Use context clues to interpret new words.

Strategy 3: Utilize vocabulary-building resources.

Strategy 4: Build a deeper knowledge of words through math application tasks and collaborative discussions.

Strategy 5: Memorize words through repetitive study such as using flashcards (digital or print) and notes.

1. First, present the shortest form of the word, referred to in this text as the "base word" in the case of academic words and some subject-specific terms. Follow the base form with other commonly used word forms (if available). Examine prefixes and suffixes and their impact on word meaning and part of speech.
2. Read the word as used in the context of the text and discuss possible meanings given context clues and word form.
3. Have students find (electronically or in print) the definition or translation of the base form and, if different, the form used in context and note these definitions in the space provided for future reference and study.
4. Gradually build a deeper knowledge of the word by having students use the word in a sentence frame, guided discussion, and an original sentence within a mathematical context.

Sentence Frame:	*A* **sphere** *is a ____________________ figure that ____________________.*
Guided Discussion:	*What are some examples of* **spheres** *in real life?*
Original Math Sentence:	____

Encourage students to use these words in math applications and collaborative discussions such as the task described in Bridging Problem Solving Strategy 4.

5. The high volume of mathematical terminology requires repeat exposure to the words over time. Word walls, intentionally including the words in questions to students and when eliciting responses from them, and explicit reminders to use the vocabulary in verbal tasks provide built-in reinforcement. However, this is often not enough so it is important that students learn ways to study words independently. Flashcards or websites that offer repetitive vocabulary practice are excellent ways for students to do this. Students may also use their notes, however, they will need to do repetitive activities, similar to flashcard practice, and not simply read and reread their notes.

➢ Bridging Math Application

Strategy 1: ***Prepare for math applications by identifying the problem type and the problem-solving strategies and tools.***

Strategy 2: ***Organize the problem using visual, symbolic, and written representations.***

Strategy 3: ***Overcome barriers to problem solving using math models, language and structural analysis, and resources.***

Strategy 4: ***Demonstrate and defend problem-solving application and mathematical reasoning through reverse problem solving, mental mathematics, visual representations, and peer discussions.***

1. Each part of this lesson lends itself to a variety of math application tasks that allow students to synthesize, apply, or extend their mathematical knowledge and skills. Whichever math application task you choose, be sure to orient students to the problem type and the problem-solving strategies and tools they may utilize. The following is an example of student directions for a math application task that synthesizes the concepts developed in this lesson.

Yard Landscaping

Complete the following task in a small group.

You are a landscaping project manager. A client comes to you with magazine pictures of 5 different improvements she wants to add to her backyard. You must figure out how to fit each of the improvements into a backyard space that is 50 feet wide and 100 feet long in a way that is both practical and beautiful. First, create a scale drawing of how to assemble the improvements in the backyard. Make sure to leave room for the lawn. Then, determine the dimensions of each of the improvements. Finally, calculate the amount of materials needed to complete each of the improvements.

Improvements:

A wood fence around the entire yard (use 6-foot, pre-made fence sections)

A rectangular pathway (use 1½-square-foot pavers)

A round patio for a fire pit (use special pavers for creating circular patios sold by the sq ft)

A rectangular concrete slab (4 inches deep) for a patio for dining area (use concrete sold by the cubic yard)

A sculpture of a concrete column with a concrete sphere on top (use concrete sold by the cubic yard)

Before engaging in problem solving, have students analyze the directions to determine what is being asked of them and use this information to determine which strategies to use to complete the task. The following example provides a glimpse of how this may look.

Directions	**Operation or Action** *Signal Words*	**Strategies and Tools**
First, create a drawing of how to assemble the improvements into the backyard.	measure, draw, label, create scale	Create a measurement scale; draw the yard to scale and label the length and width; put the improvements in where they look good
Then, determine the dimensions of each of the improvements.		

2. Students will need to organize how to complete the above task. Their analysis of the directions is a good start. They will also need a step-by-step plan for calculating the dimensions of each of the improvements and determining the materials they will need. A graphic organizer that outlines the steps they need to take and gives them a place to record their answers is a great place to start. This may take a lot of guidance to lead them toward an effective plan. For students who struggle with organizing, provide them a template but gradually release responsibility as they get more exposure to project planning.

Improvement	Dimensions	Materials Needed
A pathway with rectangular pavers		
A round patio for a fire pit		
A rectangular concrete slab (4 inches deep) for a patio for dining		
A sculpture of a concrete column with a concrete sphere on top		

3. Students can collaborate with others to organize and complete the project. They can continue to use the geometry models they created in section 1 of this unit to guide them in calculating perimeter, area, and volume.
4. Once each group of students completes their design and calculations, have groups present their plans to the "client." The client will be the classroom and students will need to choose which plan is the best and most accurate for completing the project. Students should evaluate the designs not only for their beauty but mostly for their mathematical accuracy.

➤ Assessment & Next Steps

Students should complete the practice activities included in each *Math Sense 2: Focus on Problem Solving* lesson. Evaluate which learning goals were not met and remediate by using other resources, such as those identified in the Bridging Knowledge section. Upon successful completion, continue to the next unit.

Unit 5

CONNECTING ALGEBRA AND GEOMETRY

Skills-Based Questions:

1. What is a coordinate plane and how do you find the coordinates of a point on a coordinate plane? *(Part 1)*
2. What are intercepts and how do you use them to find unknown intercepts needed to graph a linear equation? *(Part 2)*
3. What is slope and how do you calculate it given a graphed line or the slope formula? *(Part 3)*
4. What is the relationship between the slopes of parallel lines? Perpendicular lines? *(Part 4)*
5. How do you write the equation of a line? *(Part 5)*
6. How do you write the equation of a line when you cannot identify the y-intercept? *(Part 6)*
7. How do you find the distance between points on a line including non-vertical and non-horizontal lines? *(Part 7)*

Math Sense 2: Focus on Problem Solving: Part 1, p. 144; Part 2, p. 148; Part 3, p. 150; Part 4, p. 152; Part 5, p. 156; Part 6, p. 158; Part 7, p. 160

	Learning Goals:	GED
Knowledge Goals:	1. Describe a coordinate plane and how to find coordinates on the plane. *(Part 1)*	A.5.a
	2. Describe how to find unknown intercepts in a linear equation and use them to graph a line. *(Part 2)*	A.5.b
	3. Explain how to calculate slope using a graphed line or the slope formula. *(Part 3)*	A.5.d
	4. Describe the slope of parallel lines and of perpendicular lines and how you can use slope to determine if lines are parallel or perpendicular. *(Part 4)*	A.6.a
	5. Explain how to write and graph the equation of a line. *(Part 5)*	A.6.b
	6. Explain how to find the equation of a line when y-intercept is unknown. *(Part 6)*	A.6.c
	7. Describe how to find the distance between points on a line including non-vertical and -horizontal lines. *(Part 7)*	Q.4.e
Problem-Solving Goals:	1. Find coordinates on a coordinate plane. *(Part 1)*	A.5.a
	2. Use intercepts to find unknown intercepts of an equation; use the intercepts to graph a line. *(Part 2)*	A.5.b
	3. Calculate slope using a graphed line or the slope formula. *(Part 3)*	A.5.d
	4. Calculate slope to determine if lines are parallel or perpendicular. *(Part 4)*	A.6.a
	5. Write the equation of a line using the slope-intercept form. *(Part 5)*	A.6.b
	6. Use the point-slope form to write and graph the equation of a line. *(Part 6)*	A.6.c
	7. Find the distance between points on a line non-vertical and non-horizontal lines. *(Part 7)*	Q.4.e
Vocabulary Goals:	1. Define key mathematical terms.	
	2. Determine the meaning of unknown vocabulary using context clues, word forms, and parts of speech.	
	3. Apply new vocabulary to mathematical tasks and discussions.	
Math Application Goals:	1. Apply knowledge of slope and linear equations to solve real-life math problems. *(Parts 1–7)*	A.5.a–5.d
	2. Defend math applications and reasoning to others. *(Parts 1–7)*	A.6.a–6.c

Sample Instructional Support Strategies

➢ Bridging Knowledge

Strategy 1: ***Develop and connect background knowledge, skills, and conceptual understanding to new knowledge.***

Strategy 2: ***Use guiding questions to make connections beyond the lesson to broader math applications.***

Strategy 3: ***Use problem-solving strategies to develop, monitor, and synthesize conceptual understanding and fluency.*** *(See also Bridging Problem Solving)*

Strategy 4: ***Extend problem-solving skills and mathematical reasoning to broader math applications in life and work.*** *(See Bridging Math Application)*

1. Evaluate students' knowledge of the following mathematical skills. Utilize the chart below to develop student content knowledge as necessary.

Unit 5 **Connecting Algebra and Geometry**	**Part 1** The Coordinate Plane	**Part 2** Use Intercepts to Graph a Line	**Part 3** Slope: Rise Over Run	**Part 4** Parallel and Perpendicular Lines	**Part 5** Write the Equation of a Line	**Part 6** Use Point-Slope Form	**Part 7** Distance Between Points
Core Skills in Mathematics	Unit 5, Lesson 2: Using a Coordinate Plane (p. 89)						
Scoreboost Mathematics: Algebraic Reasoning	Plot Points on Coordinate Grids (p. 34)		Use the Formula to Find Slope (p. 36)		Use the Formula to Find the Equation of a Line (p. 38)		
Pre-HSE Workbook: Math 1	Using a Coordinate Plane (p. 10)	Graphing Linear Equations (p. 22)	Finding the Slope of a Line (p. 24)		Writing Equations for Lines (p. 26)		

The unit preview for each unit in the *Math Sense 2: Focus on Problem Solving* provides a list of contexts examples in which to apply the target math concepts as well as questions that connect to students' prior knowledge and experience (for Unit 5, p. 143). These questions also provide a bridge to broader math applications in life and work.

2. Each unit also provides specialized lessons that focus on problem solving, using tools, and test taking techniques using the math skills taught in the lesson. Utilize these lessons to build student knowledge in these areas.

Unit 5	**Content**	**Page**
Problem Solver	Make a Table to Graph an Equation Problem Solving with Slope Geometric Figures on the Coordinate Plane	146 154 162
Tools	NA	
Test Taker	NA	

➤ Bridging Problem Solving

Strategy 1: ***Preview the problem to determine problem-solving strategies and tools and predict general solutions.***

Strategy 2: ***Develop conceptual understanding of mathematical problems using visual representations, think-alouds, and collaboration.***

Strategy 3: ***Overcome barriers to problem solving using math models, language and structural analysis, and resources.***

Strategy 4: ***Demonstrate and defend problem solving and mathematical reasoning through reverse problem solving, mental mathematics, visual representations, and peer discussions.***

1. For each part of Unit 5, orient students to features of each mathematical concept such as the symbols, language, and structure. Students should identify the math language and symbols to determine what is being asked of them and use this information to determine which strategies to use to complete the task. Although it is not necessary for students to fill out a graphic organizer for each math problem they attempt, completing the following graphic organizer is helpful for annotating and reviewing math concepts and choosing appropriate strategies to complete tasks. The following is an example of how a student (with guidance) might fill out this graphic organizer:

	Symbols	**Words**	**Operation or Action**	**Structure**	**Strategies/Tools**
Part 1 Coordinate Plane	(x, y) – ↕↔	rectangular coordinate plane, *x*-axis, *y*-axis, origin, *x*-coordinate, *y*-coordinate, ordered pairs	find, plot	perpendicular lines make 4 quadrants: points to left of origin are positive, to the right negative; above origin are positive, below negative	line up point with *x*-axis to find *x*-coordinate; line up point with *y*-axis to find *y*-coordinate
Part 2 Use Intercepts to Graph a Line	(x, y) –	*y*-intercept, *x*-intercept, linear equation	substitute, find, graph, calculate	the *y*-intercept is 0 when a line crosses the *x*-axis, the *x*-intercept is 0 when a line crosses the *y*-intercept	substitute 0 in for the *x*-value to find the *y*-intercept; substitute 0 in for the *y*-value to find the *x*-intercept
Part 3 Slope	(x_1, y_1) $m\ \frac{\blacksquare}{\blacksquare}$	slope, rise, run, coordinate plane	count, subtract, calculate	use the slope formula on the GED math formula sheet	choose one point for (x_1, y_1) and another point for (x_2, y_2); substitute the values into the slope formula, solve
Part 4 Parallel and Perpendicular Lines	(x_1, y_1) $m\ \frac{\blacksquare}{\blacksquare}$	parallel, perpendicular, slope, rise, fall, positive slope, negative slope, negative reciprocal	find, subtract	the slope of one perpendicular line is the negative reciprocal of the other	use the slope formula to find the slope of lines; compare their slopes: same = parallel; negative reciprocal = perpendicular
Part 5 Equation of a Line	(x_1, y_1) $-\ m\ \frac{\blacksquare}{\blacksquare}$	equation, slope, *y*-intercept, slope-intercept, *y*-axis	write, substitute, find, graph, plot	$y = mx + b$; m = slope, b = *y*–value of *y*-intercept	use the slope-intercept form to find the equation of a line
Part 6 Point-Slope Form	(x_1, y_1) $-\ m\ \frac{\blacksquare}{\blacksquare}$	point-slope form, slope, *y*-intercept	write, substitute, find, graph,	$y–y_1 = m(x–x_1)$, m = slope, (x_1, y_1) is a point on the line	use point-slope form when you cannot find the *y*-intercept
Part 7 Distance Between Points	+ – = ||	horizontal, vertical, *y*-axis, *x*-axis, absolute value, square, radical	count, add, subtract, find, square, calculate	vertical and horizontal lines: count units between points; hypotenuse: use Pythagorean Theorem	find distance of vertical and horizontal lines to make a right angle; use Pythagorean Theorem to find the distance of the hypotenuse

Expanding algebraic thinking requires further exploration with math concepts in concrete ways. Here is an activity that allows students to explore the concepts outlined in Unit 5.

Coordinate Bingo:

Create a coordinate grid that goes from +2 to –2 on both the x and y axis. This will be the Bingo grid. Create "call-out" cards by writing all x coordinates and all y coordinates on small cards [$x = -2, x = -1, x = 0, x = 1, x = 2; y = -2, y = -1, y = 0, y = 1, y = 2$]. Put x coordinates in one envelope and y coordinates in another. Choose one x coordinate and one y coordinate to call out to students. Students plot a point for each pair of coordinates you call out. (Keep track of the points as well to check Bingos.) Replace cards in envelopes and draw again. Continue to call out pairs of coordinates until students have correctly plotted five points across, down, or diagonally. Of course, it is possible for all students to get a Bingo at the same time, as they have identical grids and hear the identical coordinates. However, keeping up a fast pace allows only students who are focused and understand how to plot points to get the first Bingo.

Extension:

Have students play a second round of Bingo but this time have them plot 4 coordinates (2 pairs of coordinates) and connect them with a line. This time, students will need to create line segments that go entirely across, down, or diagonally. Later, you can use their Bingo grid to create linear equations.

2. Once again, there is an overwhelming amount of mathematical vocabulary students must become familiar with. Have students continue to reference the vocabulary for this unit on p. 167. Encourage students to use the correct vocabulary whenever possible, eliciting vocabulary from them on an ongoing basis rather than supplying it.
3. When students complete a Bingo in the above activity, they must verify their points by saying them out loud to the teacher. The teacher can then check their points with the record he/she has been keeping.

➤ Bridging Vocabulary

Strategy 1: Identify the component parts and usage of new words to interpret their meanings.

Strategy 2: Use context clues to interpret new words.

Strategy 3: Utilize vocabulary-building resources.

Strategy 4: Build a deeper knowledge of words through math application tasks and collaborative discussions.

Strategy 5: Memorize words through repetitive study such as using flashcards (digital or print) and notes.

1. First, present the shortest form of the word, referred to in this text as the "base word", in the case of academic words and some subject-specific terms. Follow the base form with other commonly used word forms (if available). Examine prefixes and suffixes and their impact on word meaning and part of speech.
2. Read the word as used in the context of the text and discuss possible meanings given context clues and word form.
3. Have students find (electronically or in print) the definition or translation of the base form and, if different, the form used in context and note these definitions in the space provided for future reference and study.
4. Gradually build a deeper knowledge of the word by having students use the word in 1) a sentence frame, 2) guided discussion, and 3) an original sentence within a mathematical context.

Sentence Frame:	*Knowing the* ***slope*** *is important for* ____________ *because* ____________.
Guided Discussion:	*What are some examples of* ***slope*** *in real life?*
Original Math Sentence:	____________

Encourage students to use these words in math applications and collaborative discussions such as the task described in Bridging Problem Solving Strategy 4.

5. The high volume of mathematical terminology requires repeat exposure to the words over time. Word walls, intentionally including the words in questions to students and when eliciting responses from them, and explicit reminders to use the vocabulary in verbal tasks provide built-in reinforcement. However, this is often not enough so it is important that students learn ways to study words independently. Flashcards or websites that offer repetitive vocabulary practice are excellent ways for students to do this. Students may also use their notes, however, they will need to do repetitive activities, similar to flashcard practice, and not simply read and reread their notes.

➤ Bridging Math Application

Strategy 1: ***Prepare for math applications by identifying the problem type and the problem-solving strategies and tools.***

Strategy 2: ***Organize the problem using visual, symbolic, and written representations.***

Strategy 3: ***Overcome barriers to problem solving using math models, language and structural analysis, and resources.***

Strategy 4: ***Demonstrate and defend problem-solving application and mathematical reasoning through reverse problem solving, mental mathematics, visual representations, and peer discussions.***

1. Each part of this lesson lends itself to a variety of math application tasks that allow students to synthesize, apply, or extend their mathematical knowledge and skills. Whichever math application task you choose, be sure to orient students to the problem type and the problem-solving strategies and tools they may utilize. The following is an example of student directions for a math application task that synthesizes the concepts developed in this lesson.

Housing Market Trends

Research housing prices during the housing market "bubble" from 2002–2005 and the housing market "crash" from 2005–2008. Your research will have to contain the average housing prices for each year in the ranges above. Create a graph that shows these trends in the housing market. Label the *y*-axis with the dollar amounts in $20,000 increments starting at $100,000. Label the *x*-axis with the years starting at 2002. Plot a point for each average housing price per year. Find the average slope of the three years of increases and the average slope of the three years of decreases. (You may do this mathematically or draw a line that keeps an equal number of points on either side of it.) Which slope is greater? The increase or the decrease? Which slope has the greater distance between the beginning point and the end point? (You will need to calculate the distance between points here.)

Presentation

Present your graph and calculations to your classmates. Using correct mathematical terminology (*y*-axis, *x*-axis, *y*-coordinate, *x*-coordinate, positive slope, negative slope, etc.), describe the following steps:

1. How you plotted the points
2. How you calculated the slope
3. How you determined the average (mathematically or with a line)
4. How you determined which slope was greater
5. How you calculated the distances between beginning and end points of each slope

Before engaging in problem solving, have students analyze the directions to determine what is being asked of them and use this information to determine which strategies to use to complete the task. The following example provides a glimpse of how this may look.

Directions	**Operation or Action** *Signal Words*	**Strategies and Tools**
Research housing prices during the housing market "bubble" from 2002–2005 and the housing market "crash" from 2005–2008. Your research will have to contain the average housing prices for each year in the ranges above.	Research	Use key words "housing market," "bubble," "crash," "average housing prices"
Create a graph that shows these "trends" in the housing market. Label the *y*-axis with the dollar amounts starting at $100,000. Label the *x*-axis with the years starting at 2002.		

2. Students will need to organize how to complete the above task. Their analysis of the directions is a good start. They will also need a step-by-step plan for creating the graph and making the required calculations. A step-by-step list with space for calculations may be helpful.

Average Housing Prices Graph Coordinates	**Slope Calculations**	**Distance Calculations**
2002: 2003: 2004: 2005: 2006: 2007: 2007:	Average positive slope: Average negative slope:	Positive slope: Negative slope:
	Which slope is greater?	Which distance is greater?

3. Students can collaborate with others to organize and complete the task. They can continue to use reference sheets for the necessary mathematical terminology, especially as they prepare for their presentations.

4. The task requires that students present their graphs and explain their calculations. This is an opportunity for students to defend their mathematical reasoning and calculations as well as to probe and evaluate those of others.

➢ Assessment & Next Steps

Students should complete the practice activities included in each *Math Sense 2: Focus on Problem Solving* lesson. Evaluate which learning goals were not met and remediate by using other resources, such as those identified in the Bridging Knowledge section. Upon successful completion, continue to the Simulated GED Test.

Unit 1

DATA ANALYSIS: Section 1

Skills-Based Questions:

1. How do you use tables to draw conclusions about data? *(Part 1)*
2. How do you read bar graphs and histograms to draw conclusions about data? *(Part 2)*
3. What kind of data is best displayed on a circle graph? How do we interpret that data to draw conclusions? *(Part 3)*
4. How do scatter plots show a relationship between two sets of data? How do you interpret this relationship? *(Part 4)*
5. What kind of data is best displayed on a line graph? How do we interpret that data to draw conclusions? *(Part 5)*
6. What is the mean? What does it tell you? How do you find it? *(Part 6)*
7. What are median and mode? What information do they tell you? How do you find each one? *(Part 7)*
8. When do we use weighted averages? How do we calculate them? *(Part 8)*

Math Sense 3: Focus on Analysis: Part 1, p. 18; Part 2, p. 20; Part 3, p. 22; Part 4, p. 24; Part 5, p. 26; Part 6, p. 28; Part 7, p. 30; Part 8, p. 34

Learning Goals:

	Learning Goals	GED
Knowledge Goals:	1. Explain how tables present data and allow us to draw conclusions about data. *(Part 1)*	Q.6.a
	2. Explain how bar graphs and histograms represent data and how we use them to draw conclusions about data. *(Part 2)*	Q.6.b
	3. Describe what data is best presented on a circle graph and how we draw conclusions from it. *(Part 3)*	Q.6.c
	4. Describe the kinds of relationships between two data sets that scatter plots can show. *(Part 4)*	Q.7.a
	5. Describe what data is best presented on a line graph and how we draw conclusions from it. *(Part 5)*	
	6. Explain the meaning of *mean* and describe what information it tells us and how to calculate it. *(Part 6)*	
	7. Explain the meanings of *median* and *mode*; describe what information they tell us and how we find their values. *(Part 7)*	
	8. Explain when to use weighted averages and describe how to calculate them. *(Part 8)*	
Problem-Solving Goals:	1. Read tables and draw conclusions from the data. *(Part 1)*	Q.6.a
	2. Read bar graphs and histograms and draw conclusions from their data. *(Part 2)*	Q.6.b
	3. Read circle graphs and draw conclusions from the data. *(Part 3)*	Q.6.c
	4. Determine the relationship between two data sets on a scatter plot. *(Part 4)*	Q.7.a
	5. Read line graphs and draw conclusions from the data. *(Part 5)*	
	6. Calculate the mean, given a set of values. *(Part 6)*	
	7. Find median and mode, given a set of values; determine which best shows the typical value. *(Part 7)*	
	8. Find weighted averages. *(Part 8)*	
Vocabulary Goals:	1. Define key mathematical terms.	
	2. Determine the meaning of unknown vocabulary using context clues, word forms, and parts of speech.	
	3. Apply new vocabulary to mathematical tasks and discussions.	
Math Application Goals:	1. Apply knowledge of data analysis to solve real-life math problems. *(Parts 1–8)*	Q.6.a
	2. Defend math applications and reasoning to others. *(Parts 1–8)*	Q.6.b
		Q.6.c
		Q.7.a

Sample Instructional Support Strategies

➤ Bridging Knowledge

Strategy 1: ***Develop and connect background knowledge, skills, and conceptual understanding to new knowledge.***

Strategy 2: ***Use guiding questions to make connections beyond the lesson to broader math applications.***

Strategy 3: ***Use problem-solving strategies to develop, monitor, and synthesize conceptual understanding and fluency.*** *(See also Bridging Problem Solving)*

Strategy 4: ***Extend problem-solving skills and mathematical reasoning to broader math applications in life and work.*** *(See Bridging Math Application)*

1. Evaluate students' knowledge of the following mathematical skills. Utilize the chart below to develop student content knowledge as necessary.

Unit 1 **Data Analysis, Section 1**	**Part 1** Working with Tables	**Part 2** Bar Graphs and Histograms	**Part 3** Circle Graphs	**Part 4** Scatter Plots	**Part 5** Line Graphs	**Part 6** Finding the Mean	**Part 7** Median and Mode	**Part 8** Weighted Averages
Core Skills in Mathematics	Unit 6, Lesson 4: Representing and Interpreting Data (p. 123)					Unit 7, Lesson 1: Understanding Statistical Variability (p. 131)		
Scoreboost Mathematics: Graphs, Data Analysis, Probability	Use Tables and Charts (p. 4)	Make Comparisons with Bar Graphs (p.6) Using Frequency Tables to Make Line Plots and Histograms (p. 8)	Relate Parts and Wholes with Circle Graphs (p. 12)		See Trends with Line Graphs (p. 10)	Find Range and Mean in a Data Set (p. 16)	Find Median and Mode in a Data Set (p. 18)	
Pre-HSE Workbook: Math 2		Interpreting Bar and Line Graphs (p. 38)	Interpreting Circle Graphs (p. 40)	Interpreting Scatter Plots (p. 42)	Interpreting Bar and Line Graphs (p. 38)	Finding Measures of Central Tendency (p. 36)		

2. The unit preview for each unit in the *Math Sense 3: Focus on Analysis* provides a list of context examples in which to apply the target math concepts as well as questions that connect to students' prior knowledge and experience (for Unit 1, p. 17). These questions also provide a bridge to broader math applications in life and work.

3. Each unit of *Math Sense 3* also provides specialized lessons that focus on problem solving and using tools with the math skills taught in the lesson. Utilize these lessons to build student knowledge in these areas.

Unit 1, Section 1	**Content**	**Page**
Problem Solver	Using Typical Values	32
Tools	NA	

➤ Bridging Problem Solving

Strategy 1: ***Preview the problem to determine problem-solving strategies and tools and predict general solutions.***

Strategy 2: ***Develop conceptual understanding of mathematical problems using visual representations, think-alouds, and collaboration.***

Strategy 3: ***Overcome barriers to problem solving using math models, language and structural analysis, and resources.***

Strategy 4: ***Demonstrate and defend problem solving and mathematical reasoning through reverse problem solving, mental mathematics, visual representations, and peer discussions.***

1. For Parts 1–8 of Unit 1, orient students to features of each mathematical concept such as the symbols, language, and structure. Students should identify the math language and symbols to determine what is being asked of them and use this information to determine which strategies to use to complete the task. Although it is not necessary for students to fill out a graphic organizer for each math problem they attempt, completing the following graphic organizer is helpful for annotating and reviewing math concepts and choosing appropriate strategies to complete tasks. The following is an example of how a student (with guidance) might fill out this graphic organizer:

	Symbols	**Language**	**Operation or Action**	**Structure**	**Strategies/Tools**
Part 1 Tables	. $	table, row, column, data, categories, headings, title, conclusions, increase, decrease, difference, compared to	locate, identify, compare, subtract, add	rows across, columns down, headings across first row and down first column	read title and units; locate the correct column and row, follow each until they intersect (common cell); read questions carefully to identify operation
Part 2 Bar Graphs and Histograms	% $	bar graph, data bars, categories, vertical axis, horizontal axis, title, histogram, decrease, increase, interval	find, estimate, subtract, add, compare	bar graph compares one characteristic for different categories; histogram compares two characteristics for one category	read title, headings, and units; locate the correct category or characteristic and follow along bar; identify the value on the other axis (usually y-axis); read questions carefully to identify operation
Part 3 Circle Graphs	. % $	circle graph, sections, whole, parts, percent, fraction, decimal, total, how much, how many, combined	locate, find, identify, add, subtract	circle = whole (100%, 1, or total amount); sections = part (percent, fraction, or decimal)	read the title, headings, and units; locate the correct label and its value; read questions carefully to identify operation
Part 4 Scatter Plots	. % $	scatter plot, relationship, sets of data, survey, poll, results, horizontal axis, vertical axis, correlation, positive, negative	locate, find, predict, decide	scatter plot shows relationship between two sets of data; positive correlation if both sets increase; negative correlation if one set increases the other decreases	read the title, headings, and the units; if the points make diagonal line, there is a correlation: up to the right is positive; down to the right is negative; no line means no correlation

Part 5 Line Graphs	. % $	line graph, plot, data points, changes over time, approximate, horizontal axis, vertical, axis, decrease, increase, compare, greater than, less than	locate, find, predict, subtract, compare	line graph compares two characteristics for one category, usually for changes over time; plot values where they intersect with *x* and *y*-axes values; connect points to show increase or decrease	read title, headings, and units; locate the correct category or characteristic and follow up or across to find the data point; identify the value on the other axis; examine the line to see the increase or decrease; read questions carefully to identify operation
Part 6 Mean	+ = ÷ −	typical value, central tendency, mean median, mode, average, range	find, solve, add, divide	mean = total of values divided by the number of values	follow the process for calculating the mean; set up an equation to find a missing value using the mean
Part 7 Median and Mode	+ = ÷ −	median, least to greatest, mode, count, outliers,	sort, order, add, divide, count, identify	median = middle of data set; mode = most repeated value; if no repeated value, then no mode	order data set least to greatest; find the value in middle of data set; find average if there are two middle values; identify the most repeated value to find mode
Part 8 Weighted Average	+ = ÷ − × %	weighted average, total	multiply, add, divide, calculate, find	multiply each value by its weight, add sums together, divide by number of values; percent weight = 100%, multiply by each percent, add sums	use a visual to set up the problem to find the weight of each value; follow the steps to calculate weighted averages

2. Building data analysis skills requires further exploration with math concepts in concrete ways. Here is an activity that allows students to explore the different types of data representations and the different types of data relationships they present:

Matching Data to Representation:

Elicit from students the kinds of data (data topics) we collect in everyday life. Encourage them to include data from areas of economics, employment, education, sports, etc. (Have a list of topics prepared to ensure that there is a topic appropriate for each type of data representation.) Divide students into small groups and give each group one area of data to explore. Have them decide which questions they want their data to answer and then consider what data those questions would yield. Groups will then present the questions they want their data to answer and examples of the types of data they would collect. Through class discussion, hone their questions and data sufficiently to match at least one type of data representation.

Next, introduce students to each type of data representations: bar graph, histogram, circle graph, line graph and scatter plot. Give each group a model of each. The model should include the definition of the representation, a list of example questions one might answer using this representation, and an example of a completed representation using a common, simple example. Students should explore each of these models and choose which model they believe would best represent the data they would collect. Finally, have students present their choice to the class and explain why they chose the representation they did. Through class discussion, each group arrives at a representation that is appropriate for the data they wish to collect.

Here are some examples of the data topics students could explore, questions that will narrow the scope of that data, the data to be collected, and appropriate representations for that data:

Data Topics	Questions	Data to Collect	Representation
Average income	How has the average income changed in the last 10 years?	Average income/year for 10 years	Line graph
Student attendance	What is the attendance of students in the class?	Number of students/day	Histogram or circle graph
U.S. economic growth	Has there been economic growth over the last year?	% growth/month over last year	Line graph
Favorite sports	What are the top five favorite sports in the United States?	% of people who like top 5 sports	Bar graph or circle graph
Educational level and income	Does educational level help you earn more money?	People's income matched to educational level	Scatter plot
Time spent on daily activities	How much time do I spend on different daily activities?	Categories of activities, % of time in 24 hours	Circle graph

3. Collaboration is a big part of the above activity, both within small groups and with the full class. Use this collaboration as an opportunity for students to help each other with both the math concepts and the language. Encourage students to refer to their math vocabulary lists (p. 161) and to use the discourse prompts (p. 159). Furthermore, have students return to the provided math models as needed to enhance and check their understanding.

4. The above activity requires that students defend their choices both within their groups and to the whole class. Part of this defense must be the evidence they use to support their choices. Have students refer to the models provided in the above activity to support their positions. For example, students may defend their choice of a circle graph by saying: "In the model of the circle graph, it shows how you can show the parts of a whole. In our example, we are showing the different percents of 100%; the five top sports equal 100%." Here is some language they can use to help them support their choices:

The model shows ______________, and our example also ______________.

The definition of a circle graph is ______________, and our example ______________.

The example on the model shows ______________ like our example ______________.

We want to show ______________, and the definition of a line graph says ______________.

According to the example, we can use a histogram to ______________, so, a histogram is ______________.

➢ Bridging Vocabulary

Strategy 1: ***Identify the component parts and usage of new words to interpret their meanings.***

Strategy 2: ***Use context clues to interpret new words.***

Strategy 3: ***Utilize vocabulary-building resources.***

Strategy 4: ***Build a deeper knowledge of words through math application tasks and collaborative discussions.***

Strategy 5: ***Memorize words through repetitive study such as using flashcards (digital or print) and notes.***

1. First, present the shortest form of the word, referred to in this text as the "base word" in the case of academic words and some subject-specific terms. Follow the base form with other commonly used word forms (if available). Examine prefixes and suffixes and their impact on word meaning and part of speech.
2. Read the word as used in the context of the text and discuss possible meanings given context clues and word form.
3. Have students find (electronically or in print) the definition or translation of the base form and, if different, the form used in context and note these definitions in the space provided for future reference and study.
4. Gradually build a deeper knowledge of the word by having students use the word in a sentence frame, guided discussion, and an original sentence within a mathematical context.

Sentence Frame:	*We use **graphs** for many different things like showing ____________ and ____________.*
Guided Discussion:	*What are the different kinds of **graphs** and what data can you show on them?*
Original Math Sentence:	____

Encourage students to use these words in math applications and collaborative discussions such as the task described in Bridging Problem Solving Strategy 4.

5. The high volume of mathematical terminology requires repeat exposure to the words over time. Word walls, intentionally including the words in questions to students and when eliciting responses from them, and explicit reminders to use the vocabulary in verbal tasks provide built-in reinforcement. However, this is often not enough so it is important that students learn ways to study words independently. Flashcards or websites that offer repetitive vocabulary practice are excellent ways for students to do this. Students may also use their notes, however, they will need to do repetitive activities, similar to flashcard practice, and not simply read and reread their notes.

➢ Bridging Math Application

Strategy 1: ***Prepare for math applications by identifying the problem type and the problem-solving strategies and tools.***

Strategy 2: ***Organize the problem using visual, symbolic, and written representations.***

Strategy 3: ***Overcome barriers to problem solving using math models, language and structural analysis, and resources.***

Strategy 4: ***Demonstrate and defend problem-solving application and mathematical reasoning through reverse problem solving, mental mathematics, visual representations, and peer discussions.***

1. Each part of this lesson lends itself to a variety of math application tasks that allow students to synthesize, apply, or extend their mathematical knowledge and skills. Whichever math application task you choose, be sure to orient students to the problem type and the problem-solving strategies and tools they may utilize. The following is an example of student directions for a math application task that synthesizes the concepts developed in this lesson.

Data Presentations

Prepare:
In small groups, choose a data topic* you want to research in your class or program. Write questions you want your data to answer. Decide what data these questions will give you. Choose a data representation you will use to show your data.

Data Collection:
Create a survey to collect the data for this topic. Write questions to ask students in your class or program. Ask all students the questions and record their answers.

Data Analysis:
Create a data representation to show your data. Explain to others what the data means.

*Make sure the topic is not too personal. For example, students may not feel comfortable sharing their income or health conditions.

Before engaging in problem solving, have students analyze the directions to determine what is being asked of them and use this information to determine which strategies to use to complete the task. The following example provides a glimpse of how this may look.

Directions	**Operation or Action** *Signal Words*	**Strategies and Tools**
In small groups, choose a data topic you want to research in your class or program.	choose, decide, narrow	Choose a topic that is narrow enough to research, can be answered by people in the class or program, and is not too personal.
Write questions you want your data to answer.		

2. Students will need to organize how to complete the above task. Their analysis of the directions is a good start. They will also need a step-by-step plan for completing the task. Using a similar process as that used in the Bridging Problem Solving activity will help in the first phase of the project.

PREPARATION

Data Topic	Questions	Data to Collect	Representation
Favorite sports		Students' favorite sport Five top sports % of students who like each of those sports	Bar graph

DATA COLLECTION

Survey Questions:	Survey Data
What is your favorite sport?	

DATA ANALYSIS

What are the top five favorite sports of the students in class?	Data:
What is the percentage of total students who like each of those sports?	Calculation:
Create a data representation to show your data.	Notes:
Explain to others what the data means.	Notes:

3. Students can collaborate with others to organize and complete the task. They can continue to use reference sheets for the necessary mathematical terminology, especially as they prepare for their presentations.
4. The task requires that students present their graphs and explain their data. This is an opportunity for students to defend their mathematical reasoning and calculations as well as to probe and evaluate those of others. As in the Bridging Problem Solving activity, part of students' defense must be the evidence they use to support their choices. Have students refer to definitions and examples to support their positions.

➢ Assessment & Next Steps

Students should complete the practice activities included in each *Math Sense 3: Focus on Analysis* lesson. Evaluate which learning goals were not met and remediate by using other resources, such as those identified in the Bridging Knowledge section. Upon successful completion, continue to the next section of this unit.

DATA ANALYSIS: Section 2

Skills-Based Questions:

1. What is a box plot and what does it tell you about a set of data? *(Part 9)*
2. What is distribution and what information does it tell you about the data set? *(Part 10)*
3. How can you use line graphs to make predictions about future data values? *(Part 11)*
4. How do correlations show relationships between two sets of data? How do you interpret this relationship? *(Part 12)*
5. What is a sample population and how do you determine if is representative and unbiased? *(Part 13)*
6. How do you use two data sources to solve problems? *(Part 14)*

Math Sense 3: Focus on Analysis: Part 9, p. 38; Part 10, p. 40; Part 11, p. 42; Part 12, p. 44; Part 13, p. 46; Part 14, p. 48

	Learning Goals:	GED
Knowledge Goals:	1. Explain how a box plot displays information about a set of data. *(Part 9)*	Q.6.a
	2. Describe distribution and how its characteristics help us draw conclusions about its data. *(Part 10)*	Q.6.b
	3. Describe how line graphs show trends and how trends help us make predictions about future data values. *(Part 11)*	Q.6.c
	4. Describe the kinds of relationships that correlations show between two data sets. *(Part 12)*	
	5. Explain the meaning of sample population and describe how to evaluate whether a sample is representative and unbiased. *(Part 13)*	
	6. Describe how to find data from two sources and put it together to solve problems. *(Part 14)*	
Problem-Solving Goals:	1. Read box plots and draw conclusions from the data. *(Part 9)*	Q.6.a
	2. Draw conclusions about the data in a distribution by interpreting its characteristics. *(Part 10)*	Q.6.b
	3. Identify trends on a graph and predict future data values. *(Part 11)*	Q.6.c
	4. Determine the relationship between two data sets on a scatter plot. *(Part 12)*	
	5. Evaluate sample populations to see if they are representative and unbiased. *(Part 13)*	
	6. Solve problems using data from two data sources. *(Part 14)*	
Vocabulary Goals:	1. Define key mathematical terms.	
	2. Determine the meaning of unknown vocabulary using context clues, word forms, and parts of speech.	
	3. Apply new vocabulary to mathematical tasks and discussions.	
Math Application Goals:	1. Apply knowledge of data analysis to solve real-life math problems. *(Parts 9–14)*	Q.6.a
	2. Defend math applications and reasoning to others. *(Parts 9–14)*	Q.6.b
		Q.6.c

Sample Instructional Support Strategies

➢ Bridging Knowledge

Strategy 1: ***Develop and connect background knowledge, skills, and conceptual understanding to new knowledge.***

Strategy 2: ***Use guiding questions to make connections beyond the lesson to broader math applications.***

Strategy 3: ***Use problem-solving strategies to develop, monitor, and synthesize conceptual understanding and fluency.*** *(See also Bridging Problem Solving)*

Strategy 4: ***Extend problem-solving skills and mathematical reasoning to broader math applications in life and work.*** *(See Bridging Math Application)*

1. Evaluate students' knowledge of the following mathematical skills. Utilize the chart below to develop student content knowledge as necessary.

Unit 1 **Data Analysis, Section 2**	**Part 9** Reading a Box Plot	**Part 10** Distribution of Data	**Part 11** Making Predictions and Identifying Trends	**Part 12** Understanding Correlation	**Part 13** Sampling a Population	**Part 14** Using Two Data Sources
Core Skills in Mathematics	Unit 7, Lesson 2: Describing Distributions (p. 134)					
Scoreboost Mathematics: Graphs, Data Analysis, Probability	Find Range and Mean in a Data Set (p. 16) Find Median and Mode in a Data Set (p. 18)	Using Frequency Tables to Make Line Plots and Histograms (p. 8)	See Trends with Line Graphs (p. 10)			
Pre-HSE Workbook: Math 2	Finding Measures of Central Tendency (p. 36)		Interpreting Bar and Line Graphs (p. 38)	Interpreting Scatter Plots (p. 42)	Investigating Sampling Methods and Frequency Tables (p. 46)	

2. The unit preview for each unit in the *Math Sense 3: Focus on Analysis* provides a list of context examples in which to apply the target math concepts as well as questions that connect to students' prior knowledge and experience (for Unit 1, p. 17). These questions also provide a bridge to broader math applications in life and work.

3. Each unit of *Math Sense 3* also provides specialized lessons that focus on problem solving and using tools with the math skills taught in the lesson. Utilize these lessons to build student knowledge in these areas.

	Content	**Page**
Problem Solver	Drawing Conclusions	50
Tools	NA	

➢ Bridging Problem Solving

Strategy 1: ***Preview the problem to determine problem-solving strategies and tools and predict general solutions.***

Strategy 2: ***Develop conceptual understanding of mathematical problems using visual representations, think-alouds, and collaboration.***

Strategy 3: ***Overcome barriers to problem solving using math models, language and structural analysis, and resources.***

Strategy 4: ***Demonstrate and defend problem solving and mathematical reasoning through reverse problem solving, mental mathematics, visual representations, and peer discussions.***

1. For Parts 9–14 of Unit 1, orient students to features of each mathematical concept such as the symbols, language, and structure. Students should identify the math language and symbols to determine what is being asked of them and use this information to determine which strategies to use to complete the task. Although it is not necessary for students to fill out a graphic organizer for each math problem they attempt, completing the following graphic organizer is helpful for annotating and reviewing math concepts and choosing appropriate strategies to complete tasks. The following is an example of how a student (with guidance) might fill out this graphic organizer:

	Symbols	Words	Operation or Action	Structure	Strategies/Tools
Part 1 Box Plots		box plot, minimum, lower quartile, median, upper quartile, maximum, box-and-whisker-plot, difference		on a scale, left box for lower quartile, right box for upper quartile, median between boxes, line left from box to minimum, line right from box to maximum	identify the median, lower, and upper quartiles, and the minimum and maximum values; read questions carefully to determine operation
Part 2 Distribution of Data		distribution, curve, histogram, normal distribution curve, symmetric, mode, median, mean, outliers, skewed (left, right), variation (low, high), range	identify, find, determine	mean, median, mode are at the peak of a distribution curve; low outlier = left skew, high outlier = right skew; low variation = clustered, high variation = spread out; range = difference between high and low values	draw a curve over the bars in a histogram; identify characteristics (skew, variation, range, etc.) and use this to interpret data
Part 3 Predictions and Trends	. % $	trends, increase, decrease, remain steady, pattern, predictions	predict, estimate, think, recommend	trend = chance over time; increase = line moves up as it moves right, decrease = line moves down as it moves right	read the title, headings, and units; locate the correct category or characteristic and follow up or across to find the data point; identify the value on the other axis; examine the line to see the increase or decrease
Part 4 Correlation	. % $	positive correlation, negative correlation, linear correlation, line of best fit, nonlinear correlation, horizontal axis, vertical axis	locate, find, describe, predict	correlation shows relationship between two sets of data: linear correlation creates a positive or negative line, nonlinear correlation creates a curve	read the title, headings, and the units; if the points make diagonal line, there is a linear correlation; if points make a curve, there is a nonlinear correlation

Part 5 Sampling a Population		population, entire, sample, draw conclusions, representative population, bias (biased), skew, results, unbiased	identify, determine, choose, conclude	sample is part of the population; representative sample is a sample that is similar to the population you want study; unbiased is choosing a sample that will not favor one side or the other	identify the population and its characteristics, identify the sample and its characteristics; determine if the same is representative and unbiased
Part 6 Using Two Data Sources	+ − × ÷ = $\frac{\blacksquare}{\blacksquare}$	data source, how much, how many, difference, smaller, larger, greater than, less than, more, less	find, calculate, add, subtract	titles show the topic of tables and graphs, headings show the categories or characteristics, labels show the units	read the question to answer and identify the data types to collect; collect the data from two data sources and organize it in an equation

2. Building data analysis skills requires further exploration with math concepts in concrete ways. Graphs are used in nearly every area of life to show information. Often, these illustrations are created with a goal in mind, such as to sway public opinion or present something in a favorable or negative light. The fact that graphs are often misused in this way makes it imperative that our students use their skills in data analysis to evaluate the information they see on a daily basis. Doing this evaluation prior to a more in-depth look at data analysis not only helps them explore the way graphs are used to present information but also increases their motivation to understand the concepts behind data analysis. The Drawing Conclusions Problem Solver (p. 50) in *Math Sense 3: Focus on Analysis* Unit 1 provides an excellent exploration activity for students. Using this as a springboard, work with students to evaluate examples of graphs in the media.

Evaluation of Graphs in Media

Beginning with the Drawing Conclusions Problem Solver lesson on p. 50 of *Math Sense 3,* lead students through examples of how to evaluate graphs. Then provide students with graphs from mainstream media or provide newspapers and magazines that contain graphs for students to find.

In pairs or small groups, have students evaluate one of the graphs. Make sure that they read the title, the headings, and the units. Have them examine the scale. Have students answer questions such as the following:

1. Is the graph easy to read? Can you find important information quickly?
2. Does the graph fit with the type of data it is showing?
3. Does the scale make the data seem more or less important, greater or less than it is, or favor one side or another? If so, what is it about the scale that makes it seem this way? Why do you think the author made the graph this way?
4. Is there anything missing from the graph that misleads you? What is missing and why is that misleading?

Then have students either meet with other pairs (or small groups and share their evaluation or share have them it with the class.

3. Collaboration is a big part of the above activity, both within small groups and with the full class. Use this collaboration as an opportunity for students to help each other with both the math concepts and the language. Encourage students to refer to their math vocabulary lists (p. 161) and to use the discourse prompts (p. 159). Furthermore, have students return to the Drawing Conclusions Problem Solver lesson to enhance and check their understanding.

4. The previous activity requires that students defend their choices. Part of their defense will be the evidence they use to support their positions. Students must supply evidence from the graph that supports their answer to each question. Students who disagree with the answer must supply contradictory evidence to dispute the position. This particular topic involves subjectivity and may spark a lively debate. As such, make sure students follow the norms of discussion and polite disagreement.

Take turns speaking. Do not take up all the time yourself.

Listen to the person who is speaking and wait until he or she is finished before speaking.

If you disagree, use polite language to explain your disagreement AND use evidence to support your position:

- I understand what you are saying, but I think ______________ because ______________.
- You may have a point but what I believe is ______________ because ______________.
- I don't think that is correct because ______________.
- You have an interesting idea/perspective, but I believe ______________ because ______________.

➤ Bridging Vocabulary

Strategy 1: Identify the component parts and usage of new words to interpret their meanings.

Strategy 2: Use context clues to interpret new words.

Strategy 3: Utilize vocabulary-building resources.

Strategy 4: Build a deeper knowledge of words through math application tasks and collaborative discussions.

Strategy 5: Memorize words through repetitive study such as using flashcards (digital or print) and notes.

1. First, present the shortest form of the word, referred to in this text as the "base word" in the case of academic words and some subject-specific terms. Follow the base form with other commonly used word forms (if available). Examine prefixes and suffixes and their impact on word meaning and part of speech.
2. Read the word as used in the context of the text and discuss possible meanings given context clues and word form.
3. Have students find (electronically or in print) the definition or translation of the base form and, if different, the form used in context and note these definitions in the space provided for future reference and study.
4. Gradually build a deeper knowledge of the word by having students use the word in a sentence frame, guided discussion, and an original sentence within a mathematical context.

Sentence Frame:	*We use the words **maximum** and **minimum** to talk about ______________.*
Guided Discussion:	*What is the **maximum** amount of money you would spend on a car? The **minimum**? Why?*
Original Math Sentence:	______________________________

Encourage students to use these words in math applications and collaborative discussions such as the task described in Bridging Problem Solving Strategy 4.

5. The high volume of mathematical terminology requires repeat exposure to the words over time. Word walls, intentionally including the words in questions to students and when eliciting responses from them, and explicit reminders to use the vocabulary in verbal tasks provide built-in reinforcement. However, this is often not enough so it is important that students learn ways to study words independently. Flashcards or websites that offer repetitive vocabulary practice are excellent ways for students to do this. Students may also use their notes, however, they will need to do repetitive activities, similar to flashcard practice, and not simply read and reread their notes.

➤ Bridging Math Application

Strategy 1: ***Prepare for math applications by identifying the problem type and the problem-solving strategies and tools.***

Strategy 2: ***Organize the problem using visual, symbolic, and written representations.***

Strategy 3: ***Overcome barriers to problem solving using math models, language and structural analysis, and resources.***

Strategy 4: ***Demonstrate and defend problem-solving application and mathematical reasoning through reverse problem solving, mental mathematics, visual representations, and peer discussions.***

1. Each part of this lesson lends itself to a variety of math application tasks that allow students to synthesize, apply, or extend their mathematical knowledge and skills. Whichever math application task you choose, be sure to orient students to the problem type and the problem-solving strategies and tools they may utilize. The following is an example of student directions for a math application task that synthesizes the concepts developed in this lesson.

Misleading Graphs

You own a company and are going to make an advertisement to promote your company. You think using research will get people to buy your product or service. You want to create a graph in your advertisement to show this research in a favorable way to the public.

1. First, decide what your company is and what product or service you want it to provide. Second, choose the sample you will research and the data you will collect.
2. Then, create imaginary data that is either *not favorable* to your company or is *neutral.*
3. Create a graph that fairly represents the data. Make sure you use the appropriate graph type to represent your data.
4. Finally, create a new graph to make the data appear *favorable* to your company. You cannot change the data or lie about the data (this will result in an expensive fine). However, you may remove some data, change the scale or units of your graph, use creativity in your title, headings, and labels, and include visuals to enhance how your graph looks.
5. Share your new graph with a group or with the class. Can they identify what is misleading about your graph?

Before engaging in problem solving, have students analyze the directions to determine what is being asked of them and use this information to determine which strategies to use to complete the task. The following example provides a glimpse of how this may look.

Directions	**Operation or Action** *Signal Words*	**Strategies and Tools**
First, decide what your company is and what product or service you want it to provide.		
Choose the sample you will research and the data you will collect.	choose, plan	Describe the population, describe a representative sample

2. Students will need to organize how to complete the above task. Their analysis of the directions is a good start. They will also need a step-by-step plan for completing the task and somewhere to record their notes and work. Here is one example of how this may look:

Company Product/Service	**Sample Population**	**Data to Collect**	**Data (values) Collected**
Description:	Description:		
Fair Graph			
Type:	Create:		
Misleading Graph			
Type:	Changes to make:		
	Create:		

3. Students can collaborate with others to organize and complete the task. They can continue to use reference sheets for the necessary mathematical terminology, especially as they prepare to evaluate their peers graphs.

4. The task requires that students evaluate the graphs from other students to determine how they are misleading. In this way, the onus is on the audience not the presenter. Students in the audience should use what they learned about evaluating graphs to critique the graphs of others. They will need to use evidence to support their positions.

➢ Assessment & Next Steps

Students should complete the practice activities included in each *Math Sense 3: Focus on Analysis* lesson. Evaluate which learning goals were not met and remediate by using other resources, such as those identified in the Bridging Knowledge section. Upon successful completion, continue to the next unit.

Unit 2

COUNTING AND PROBABILITY

Skills-Based Questions:

1. What are permutations and how do you find them? *(Part 1)*
2. What is distribution and what information does it tell you about the data set? *(Part 2)*
3. How do you find simple probability? How do you find odds? How are they different? *(Part 3)*
4. How do you find compound probability? How are the calculations different for independent and dependent compound probabilities? *(Part 4)*
5. What is experimental probability and how is it different than simple probability? How do you use probability to predict possible outcomes? *(Part 5)*

Math Sense 3: Focus on Analysis: Part 1, p. 60; Part 2, p. 62; Part 3, p. 64; Part 4, p. 66; Part 5, p. 70

	Learning Goals:	**GED**
Knowledge Goals:	1. Explain the meaning of permutations and describe how to find them. *(Part 1)*	Q.8.a
	2. Explain the meaning of combinations and describe how to find them. *(Part 2)*	Q.8.b
	3. Describe how to find probability and odds and explain how they are different. *(Part 3)*	
	4. Describe how to find compound probability and explain how the process is different for finding independent vs. dependent compound probabilities. *(Part 4)*	
	5. Explain how experimental probability and simple probability are different; explain how to use probability to predict possible outcomes. *(Part 5)*	
Problem-Solving Goals:	1. Find the permutations of a set of values. *(Part 1)*	Q.8.a
	2. Find the combinations of a set of values; determine whether to find the permutation or the combination. *(Part 2)*	Q.8.b
	3. Find probabilities and odds. *(Part 3)*	
	4. Find independent and dependent compound probabilities. *(Part 4)*	
	5. Find experimental probability and use it to predict possible outcomes. *(Part 5)*	
Vocabulary Goals:	1. Define key mathematical terms.	
	2. Determine the meaning of unknown vocabulary using context clues, word forms, and parts of speech.	
	3. Apply new vocabulary to mathematical tasks and discussions.	
Math Application Goals:	1. Apply knowledge of probability to solve real-life math problems. *(Parts 1–5)*	Q.8.a
	2. Defend math applications and reasoning to others. *(Parts 1–5)*	Q.8.b

Sample Instructional Support Strategies

➢ Bridging Knowledge

Strategy 1: ***Develop and connect background knowledge, skills, and conceptual understanding to new knowledge.***

Strategy 2: ***Use guiding questions to make connections beyond the lesson to broader math applications.***

Strategy 3: ***Use problem-solving strategies to develop, monitor, and synthesize conceptual understanding and fluency.*** *(See also Bridging Problem Solving)*

Strategy 4: ***Extend problem-solving skills and mathematical reasoning to broader math applications in life and work.*** *(See Bridging Math Application)*

1. Evaluate students' knowledge of the following mathematical skills. Utilize the chart below to develop student content knowledge as necessary.

Unit 2 **Counting and Probability**	**Part 1** Permutations	**Part 2** Combinations	**Part 3** Simple Probability	**Part 4** Compound Probability	**Part 5** Experimental Probability
Core Skills in Mathematics					
Scoreboost Mathematics: Graphs, Data Analysis, Probability	Use Tree Diagrams with Combinations and Permutations (p. 26)		Solve Simple Probability Problems (p. 28)	Solve Compound Probability Problems (p. 30)	
Pre-HSE Workbook: Math 2			Understanding Probability (p. 48)		

2. The unit preview for each unit in the *Math Sense 3: Focus on Analysis* provides a list of context examples in which to apply the target math concepts as well as questions that connect to students' prior knowledge and experience (for Unit 2, p. 57). These questions also provide a bridge to broader math applications in life and work.
3. Each unit of *Math Sense 3* also provides specialized lessons that focus on problem solving and using tools with the math skills taught in the lesson. Utilize these lessons to build student knowledge in these areas.

Unit 2	**Content**	**Page**
Problem Solver	Counting Strategies Using Probability for Predictions	58 68
Tools	NA	

➢ Bridging Problem Solving

Strategy 1: ***Preview the problem to determine problem-solving strategies and tools and predict general solutions.***

Strategy 2: ***Develop conceptual understanding of mathematical problems using visual representations, think-alouds, and collaboration.***

Strategy 3: ***Overcome barriers to problem solving using math models, language and structural analysis, and resources.***

Strategy 4: ***Demonstrate and defend problem solving and mathematical reasoning through reverse problem solving, mental mathematics, visual representations, and peer discussions.***

1. For each part of Unit 2, orient students to features of each mathematical concept such as the symbols, language, and structure. Students should identify the math language and symbols to determine what is being asked of them and use this information to determine which strategies to use to complete the task. Although it is not necessary for students to fill out a graphic organizer for each math problem they attempt, completing the following graphic organizer is helpful for annotating and reviewing math concepts and choosing appropriate strategies to complete tasks. The following is an example of how a student (with guidance) might fill out this graphic organizer:

	Symbols	Language	Operation or Action	Structure	Strategies/Tools
Part 1 Permutations		permutations, order, total, how many, possible, possibilities, different ways	order, count, organize, multiply	count down (one less each time) the number of choices for a limited order; multiply the number of choices together to get total permutations	read question carefully to identify the limited order; count down the choices; follow the process for calculating permutations
Part 2 Combinations		combination, random, possible, possibilities, how many, different	find, calculate, count down, multiply	find permutations of the total group and multiply together; find permutations of the chosen group and multiply together; divide possibilities of the total by the chosen to get possible combinations	read the question carefully to determine if order matters; if order matters, calculate permutations (greater number of possibilities); if order doesn't matter, calculate combinations (fewer number of possibilities)
Part 3 Simple Probability	÷ – %	probability, likelihood, chance, ratio, favorable outcomes, possible outcomes, percent, odds	find, solve, divide, subtract	probability = number of favorable outcomes over total number of possible outcomes; odds = number of favorable outcomes over unfavorable outcomes	read the question carefully to determine whether to calculate probability or odds; use the correct formula find the solution; change to percent for probability, keep as ratio (in lowest terms) for odds
Part 4 Compound Probability	÷ – × %	compound probability, multiply, independent, dependent	find, solve, divide, multiply	independent compound probability means each event does not rely on the next; dependent compound probability means the first event affects the next event	for dependent compound probability, the number of favorable outcomes or possible outcomes changes for the second event
Part 5 Experimental Probability	÷ – × %	experimental probability, favorable outcomes, trials, prediction	find, solve, divide, multiply	experimental probability = favorable outcomes over number of trials; multiply total possibilities by % probability to predict a possible outcome	use the formula for experimental probability for a sample of possibilities; apply the probability to a larger sample to predict a possible outcome

2. The internet is filled with examples of how to bring permutations, combinations, and probability to life. Unit 2 of *Math Sense 3* describes an excellent exploration activity on p. 70. Using this as a springboard, work with students to explore probability.

Probability With Skittles

Permutations:

Give each student 4 candy Skittles, each of a different color. Tell students that each colored Skittle has the opportunity to receive a ranking from 1 or 2 for best flavor. How many possible permutations are there for this ranking? First, have students count to find the permutations. Then have them use the method of counting down from the total choices (4 Skittles) for each of the total available positions (rankings 1 to 2) and multiply the numbers together.

# of Skittles → $4 \times 3 = 12$

positions

Combinations:

Using the same Skittles, now have students find how many possible color combinations of 2 there are. First have students count the permutations. Then have students count down from the total choices (4 Skittles) for each chosen set (2) and multiply the numbers together. Then have them count down from the chosen set (2) and multiply the numbers together. Finally, have them divide the possibilities from the total choices by the possibilities of the chosen set.

# of Skittles → $4 \times 3 = 12 \div 2 \times 1 = 2$ $\frac{12}{2} = 6$ possibilities

chosen set chosen set

Probability:

Give students a small paper cup of a random number of randomly colored Skittles. (They can eat the ones they were using or add them to the cup.) Have students find the experimental probability of choosing a red skittle from their cup. Have students make 20 trials and record their findings and then calculate their results. Then have students count the total number of Skittles in the cup and the number of red Skittles. Did their probability predict the number or red Skittles in their cup? Finally, have them calculate the simple probability of choosing a red Skittle. How close was their experimental probability to the simple probability?

Compound Probability:

Have students find the probability of choosing 2 red Skittles in a row, if they replace the first the red Skittle after drawing it. Then have them find the probability of choosing 2 red Skittles after keeping the first red Skittle out after drawing it.

3. Make the above activity collaborative by having students check their methods and findings with a partner. They can then meet with another pair to discuss their work. Encourage students to refer to their math vocabulary lists (p. 161) and to use the discourse prompts (p. 159).

4. The above activity allows students to test their findings by using a different process to check their work. For example, first students count to find permutations and combinations and then they use the method for calculating them. Similarly, students find the experimental probability and then check their work by calculating simple probability. Follow each of these practices with a discussion that allows students to explore their findings and enhance their mathematical reasoning.

➤ Bridging Vocabulary

Strategy 1: Identify the component parts and usage of new words to interpret their meanings.

Strategy 2: Use context clues to interpret new words.

Strategy 3: Utilize vocabulary-building resources.

Strategy 4: Build a deeper knowledge of words through math application tasks and collaborative discussions.

Strategy 5: Memorize words through repetitive study such as using flashcards (digital or print) and notes.

1. First, present the shortest form of the word, referred to in this text as the "base word" in the case of academic words and some subject-specific terms. Follow the base form with other commonly used word forms (if available). Examine prefixes and suffixes and their impact on word meaning and part of speech.
2. Read the word as used in the context of the text and discuss possible meanings given context clues and word form.
3. Have students find (electronically or in print) the definition or translation of the base form and, if different, the form used in context and note these definitions in the space provided for future reference and study.

4. Gradually build deeper knowledge of the word by having students use the word in a sentence frame, guided discussion, and an original sentence within a mathematical context.

Sentence Frame:	*Calculating **probability** can help us ________________.*
Guided Discussion:	*When do you use **probability** in your life? How does it help you?*
Original Math Sentence:	____

Encourage students to use these words in math applications and collaborative discussions such as the task described in Bridging Problem Solving Strategy 4.

5. The high volume of mathematical terminology requires repeat exposure to the words over time. Word walls, intentionally including the words in questions to students and when eliciting responses from them, and explicit reminders to use the vocabulary in verbal tasks provide built-in reinforcement. However, this is often not enough so it is important that students learn ways to study words independently. Flashcards or websites that offer repetitive vocabulary practice are excellent ways for students to do this. Students may also use their notes, however, they will need to do repetitive activities, similar to flashcard practice, and not simply read and reread their notes.

➢ Bridging Math Application

Strategy 1: Prepare for math applications by identifying the problem type and the problem-solving strategies and tools.

Strategy 2: Organize the problem using visual, symbolic, and written representations.

Strategy 3: Overcome barriers to problem solving using math models, language and structural analysis, and resources.

Strategy 4: Demonstrate and defend problem-solving application and mathematical reasoning through reverse problem solving, mental mathematics, visual representations, and peer discussions.

1. Each part of this lesson lends itself to a variety of math application tasks that allow students to synthesize, apply, or extend their mathematical knowledge and skills. Whichever math application task you choose, be sure to orient students to the problem type and the problem-solving strategies and tools they may utilize. The following is an example of student directions for a math application task that synthesizes the concepts developed in this lesson.

Counting Cards

Counting cards is not allowed in casinos. Why do you think that is? Research what it means to count cards. Imagine you are playing a game of cards with four people. Each person is dealt 5 cards to start the game. Determine the probability that you could start the game with each of the following cards (*Hint:* there are 52 cards in a deck and four of each type of card in a deck):

- 1 Ace
- 1 King
- 1 Queen

You are very lucky and are dealt 1 Ace, 1 King, and 1 Queen in your hand of cards. You also get a 2 and a 5. Now each person can remove the cards they do not like from their hand: one person removes 3 cards, another removes 2 cards, another removes 1 card, and you remove 2 cards (the 2 and 5). The dealer replaces each person's removed cards with *new* cards from the deck. What is the probability that you will receive another Ace, King, or Queen?

If you continued to calculate the probability (something similar to counting cards) throughout the game, do you think you would make better choices in the game? Why or why not?

Before engaging in problem solving, have students analyze the directions to determine what is being asked of them and use this information to determine which strategies to use to complete the task. The following example provides a glimpse of how this may look.

Directions	**Operation or Action** *Signal Words*	**Strategies and Tools**
Determine the probability that you could start the game with each of the following cards. (*Hint:* There are 52 cards in a deck and four of each type of card in a deck.): • 1 Ace • 1 King • 1 Queen	determine, count, subtract	Calculate the probability of getting each of the cards given five cards; multiply the probabilities together to find the probability of getting all of the cards at one time
What is the probability that you will receive another Ace, King, or Queen?		

2. Students will need to organize how to complete the above task. Their analysis of the directions is a good start. They will also need a step-by-step plan for completing the task and somewhere to record their notes and work. Here is one example of how this may look:

	Starting Hand		**Hand with Replaced Cards**
# of Favorable Outcomes	4	# of Favorable Outcomes	
# of Possible Outcomes	52	# of Possible Outcomes	
Independent Compounded Probability		Dependent Compounded Probability	

3. Students can collaborate with others to organize and complete the task. They can continue to use reference sheets for the necessary mathematical terminology, especially as they prepare to answer the last question: "If you continued to calculate the probability (something similar to counting cards) throughout the game, do you think you would make better choices in the game? Why or why not?"

4. Like the other lessons in *Bridging*, giving students the opportunity to share their calculations with others gives them the opportunity to defend their reasoning and skills and question those of others. When this becomes a norm in the classroom, students invest more in their work knowing that they will be accountable for explaining it. They also develop skills in negotiating using mathematical terminology which improves both their comprehension of the terms and their development of the skills behind them.

➢ Assessment & Next Steps

Students should complete the practice activities included in each *Math Sense 3: Focus on Analysis* lesson. Evaluate which learning goals were not met and remediate by using other resources, such as those identified in the Bridging Knowledge section. Upon successful completion, continue to the next unit.

Unit 3

SYSTEMS OF EQUATIONS AND INEQUALITIES

Skills-Based Questions:

1. What is a system of equations and how do you find the solution to a system by graphing? *(Part 1)*
2. How do you find a solution to a system of equations using substitution? *(Part 2)*
3. How do you find a solution to a system of equations using elimination? *(Part 3)*
4. What is a linear equality and how do you graph it? How do you find its solution on a plane? *(Part 4)*
5. What is a system of inequalities and how do you find the range of solutions to the system by graphing? *(Part 5)*

Math Sense 3: Focus on Analysis: Part 1, p. 76; Part 2, p. 78; Part 3, p. 80; Part 4, p. 84; Part 5, p. 86

	Learning Goals:	**GED**
Knowledge Goals:	1. Explain the meaning of a system of equations and describe how to find the solution to a system by graphing. *(Part 1)*	A.2.b
	2. Explain how to find the solution to a system of equations by using substitution. *(Part 2)*	A.2.c
	3. Explain how to find the solution to a system of equations by using elimination. *(Part 3)*	A.2.d
	4. Explain what a linear equality is and how it is graphed; describe how to find its solution on a coordinate plane. *(Part 4)*	A.3.b
	5. Explain the meaning of a system of inequalities and describe how to find the range of solutions to the system by graphing. *(Part 5)*	A.3.c A.5.d
Problem-Solving Goals:	1. Solve a system of equations by graphing. *(Part 1)*	A.2.b
	2. Solve a system of equations using substitution. *(Part 2)*	A.2.c
	3. Solve a system of equations using elimination. *(Part 3)*	A.2.d
	4. Solve two-variable linear inequalities by graphing. *(Part 4)*	A.3.b
	5. Find the range of solutions to a system of inequalities by graphing. *(Part 5)*	A.3.c
Vocabulary Goals:	1. Define key mathematical terms.	
	2. Determine the meaning of unknown vocabulary using context clues, word forms, and parts of speech.	
	3. Apply new vocabulary to mathematical tasks and discussions.	
Math Application Goals:	1. Apply knowledge of systems of equations to solve real-life math problems. *(Parts 1–5)*	A.2.b
	2. Defend math applications and reasoning to others. *(Parts 1–5)*	A.2.c
		A.2.d
		A.3.b
		A.3.c

Sample Instructional Support Strategies

➤ Bridging Knowledge

Strategy 1: ***Develop and connect background knowledge, skills, and conceptual understanding to new knowledge.***

Strategy 2: ***Use guiding questions to make connections beyond the lesson to broader math applications.***

Strategy 3: ***Use problem-solving strategies to develop, monitor, and synthesize conceptual understanding and fluency.*** *(See also Bridging Problem Solving)*

Strategy 4: ***Extend problem-solving skills and mathematical reasoning to broader math applications in life and work.*** *(See Bridging Math Application)*

1. Evaluate students' knowledge of the following mathematical skills. Utilize the chart below to develop student content knowledge as necessary.

Unit 3 **Systems of Equations and Inequalities**	**Part 1** Systems of Equations	**Part 2** Solve Systems Using Substitution	**Part 3** Solve Systems Using Elimination	**Part 4** Graph Linear Inequalities	**Part 5** Solve Systems of Linear Inequalities
Core Skills in Mathematics	Unit 4, Lesson 4: Graphing Equations and Inequalities (p. 72)				
Scoreboost Mathematics: Algebraic Reasoning	Fill-in-the-Blank Equations and Inequality Questions (p. 22)				
Pre-HSE Workbook: Math 2	Solving Systems of Equations (p. 32)			Graphing Linear Inequalities (p. 30)	

2. The unit preview for each unit in the *Math Sense 3: Focus on Analysis* provides a list of context examples in which to apply the target math concepts as well as questions that connect to students' prior knowledge and experience (for Unit 3, p. 75). These questions also provide a bridge to broader math applications in life and work.
3. Each unit of *Math Sense 3* also provides specialized lessons that focus on problem solving and using tools with the math skills taught in the lesson. Utilize these lessons to build student knowledge in these areas.

Unit 3	**Content**	**Page**
Problem Solver	Write Systems to Solve Word Problems Interpret Graphs of Systems	82 88
Tools	NA	

➤ Bridging Problem Solving

Strategy 1: ***Preview the problem to determine problem-solving strategies and tools and predict general solutions.***

Strategy 2: ***Develop conceptual understanding of mathematical problems using visual representations, think-alouds, and collaboration.***

Strategy 3: ***Overcome barriers to problem solving using math models, language and structural analysis, and resources.***

Strategy 4: ***Demonstrate and defend problem solving and mathematical reasoning through reverse problem solving, mental mathematics, visual representations, and peer discussions.***

1. For each part of Unit 3, orient students to features of each mathematical concept such as the symbols, language, and structure. Students should identify the math language and symbols to determine what is being asked of them and use this information to determine which strategies to use to complete the task. Although it is not necessary for students to fill out a graphic organizer for each math problem they attempt, completing the following graphic organizer is helpful for annotating and reviewing math concepts and choosing appropriate strategies to complete tasks. The following is an example of how a student (with guidance) might fill out this graphic organizer:

	Symbols	**Language**	**Operation or Action**	**Structure**	**Strategies/Tools**
Part 1 Systems of Equations	$+ \; - \; () \; \div \; = \; x \; y \; \{\}$	system of equations, variables, brace, graph, *x*-axis, *y*-axis, *x*-coordinate, *y*-coordinate	solve, graph, plot	two or more equations with the 2 variables (*x*, *y*); variables have the same values in each equation	choose values on or near zero; substitute into each equation; plot ordered pairs; find intersection to find the solution
Part 2 Using Substitution	$+ \; - \; () \; \div \; = \; x \; y \; \{\}$	system of equations, variables, substitution method, isolate	substitute, solve, rewrite, isolate	an expression that is equal to a variable in one equation may be substituted for that variable in the other equation	isolate a variable that is easy to isolate; use its equal expression in the other equation
Part 3 Using Elimination	$+ \; - \; () \; \div \; = \; x \; y \; \{\}$	elimination, combine, like terms	eliminate, combine, multiply	each variable has the same value in each of the equations in the system; solving for one variable in one equation makes it possible to solve for the other in the other equation	when terms with the same variable in each equation can be added together to equal zero, combine equations; or, use the least common multiple for a variable's coefficients to set up equations for combining
Part 4 Graph Linear Equalities	$\geq \; \leq \; > \; <$	inequality, range of solutions, shaded portion, variables, solution	graph, plot, shade	linear inequality with one variable can be graphed on a line; inequality with two variables can be graphed on a coordinate plane; solid line for $\geq$ $\leq$; dotted line for $>$ $<$	graph inequalities like equations; an inequality divides the plane in two: one side has solutions the other does not; choose coordinates and use in the equations to find the region that contains solutions
Part 5 Systems of Linear Equalities		system of inequalities, range of solutions, shading, overlap	graph, plot, shade	a system of inequalities has more than one inequality; when graphed, the system's range of solutions is in the overlapping region	graph inequalities like equations; choose coordinates from each overlapping region to substitute into one of the equations until you find coordinates that work; shade this region

2. Unit 3 of provides multiple ways to solve systems of equations. An exploration activity that allows students to try these different ways will help students develop a larger arsenal of strategies to apply to their problem solving. Here is one example of how students may explore these methods.

Solving Systems of Equations

- First, review with students how to solve linear equations. Then provide a simple example of a system of equations to build conceptual understanding. The Problem Solver lesson Write Systems to Solve Word Problems on page 82 in *Math Sense 3* presents several such examples.
- Next, using these examples, have students practice each of the methods presented in Unit 3—graphing, substitution, and elimination. Students will begin to see how one method works better for certain equations over other methods.
- Finally, begin working through the exercises in Unit 3.

3. Make the above activity collaborative by having students check their methods and calculations with a partner. They can then meet with another pair to discuss their work. Encourage students to refer to their math vocabulary lists (p. 161) and to use the discourse prompts (p. 159).
4. Solving for systems of equations has a built-in mechanism for backwards problem solving. Unit 3 uses this strategy throughout the lessons to help students check their work. Another way to do this is for students to pass their solutions to a partner and the partner checks them by substituting the solved valued into each of the equations. If there is a discrepancy, then students must discuss the methods and calculations used to determine where the problem lies and how to come up with an accurate solution.

➢ Bridging Vocabulary

Strategy 1: Identify the component parts and usage of new words to interpret their meanings.

Strategy 2: Use context clues to interpret new words.

Strategy 3: Utilize vocabulary-building resources.

Strategy 4: Build a deeper knowledge of words through math application tasks and collaborative discussions.

Strategy 5: Memorize words through repetitive study such as using flashcards (digital or print) and notes.

1. First, present the shortest form of the word, referred to in this text as the "base word" in the case of academic words and some subject-specific terms. Follow the base form with other commonly used word forms (if available). Examine prefixes and suffixes and their impact on word meaning and part of speech.
2. Read the word as used in the context of the text and discuss possible meanings given context clues and word form.
3. Have students find (electronically or in print) the definition or translation of the base form and, if different, the form used in context and note these definitions in the space provided for future reference and study.
4. Gradually build a deeper knowledge of the word by having students use the word in a sentence frame, guided discussion, and an original sentence within a mathematical context.

Sentence Frame:	*In a **system of equations**, a variable* ____________________.
Guided Discussion:	*When do we use **systems of equations** in our lives? How does this help us?*
Original Math Sentence:	____________________

Encourage students to use these words in math applications and collaborative discussions such as the task described in Bridging Problem Solving Strategy 4.

5. The high volume of mathematical terminology requires repeat exposure to the words over time. Word walls, intentionally including the words in questions to students and when eliciting responses from them, and explicit reminders to use the vocabulary in verbal tasks provide built-in reinforcement. However, this is often not enough so it is important that students learn ways to study words independently. Flashcards or websites that offer repetitive vocabulary practice are excellent ways for students to do this. Students may also use their notes, however, they will need to do repetitive activities, similar to flashcard practice, and not simply read and reread their notes.

➤ Bridging Math Application

Strategy 1: ***Prepare for math applications by identifying the problem type and the problem-solving strategies and tools.***

Strategy 2: ***Organize the problem using visual, symbolic, and written representations.***

Strategy 3: ***Overcome barriers to problem solving using math models, language and structural analysis, and resources.***

Strategy 4: ***Demonstrate and defend problem-solving application and mathematical reasoning through reverse problem solving, mental mathematics, visual representations, and peer discussions.***

1. Each part of this lesson lends itself to a variety of math application tasks that allow students to synthesize, apply, or extend their mathematical knowledge and skills. Whichever math application task you choose, be sure to orient students to the problem type and the problem-solving strategies and tools they may utilize. The following is an example of student directions for a math application task that synthesizes the concepts developed in this lesson.

Comparing Cell Phone Plans

You are shopping for a cell phone plan and have three choices.

The first plan charges $50 a month for using up to 10 gigabytes of data. For each gigabyte over the initial 10 gigabyte limit, it will cost an additional $5 charge per gigabyte on top of the initial rate per gigabyte ($10/gigabyte).

The second plan offers each gigabyte of data for $5 a month up to 8 gigabytes plus an additional $1 ($6/gigabyte) for each gigabyte over 8 gigabytes.

The third plan provides 5 gigabytes of data for $25 a month and each additional gigabyte costing an extra $2 ($7/gigabyte) with no limit to the number of gigabytes that may be purchased.

Given the amount of data you usually use each month, which is the most cost effective plan for you? (*Hint:* You will need to estimate the actual amount of data you use in a month in order to compare the three plans to find the one that is most cost effective for you.)

Teacher's Note: Answers will vary depending on each student's estimated data usage per month.

Before engaging in problem solving, have students analyze the directions to determine what is being asked of them and use this information to determine which strategies to use to complete the task. The following example provides a glimpse of how this may look.

Directions	**Operation or Action** *Signal Words*	**Strategies and Tools**
The first plan offers 10 gigabytes of data for $50 a month. Each gigabyte past 10 is $10	graph, substitute, or eliminate	Figure out what *x* and *y* represent; set up an equation that equals $50.
The second plan costs $5 for each gigabyte of data a month up to 8 gigabytes and $6 for each gigabyte beyond that.		

2. Students will need to organize how to complete the above task. Their analysis of the directions is a good start. They will also need a step-by-step plan for completing the task and somewhere to record their notes and work. Here is one example of how this may look:

	Plan 1	**Plan 2**	**Plan 3**
Known cost	\$50		
Number of gigabytes (x)	10		
Cost per gigabyte (y)	y		
Equation	$10 \times y = \$50$		
Calculations			

3. Students can collaborate with others to organize and complete the task. They can continue to use reference sheets for the necessary mathematical terminology, especially as they prepare to explain their reasoning for choosing the plan they chose.

4. Like the other lessons in *Bridging*, giving students the opportunity to share their calculations with others gives them the opportunity to defend their reasoning and skills and question those of others. When this becomes a norm in the classroom, students invest more in their work knowing that they will be accountable for explaining it. They also develop skills in negotiating using mathematical terminology, which improves both their comprehension of the terms and their development of the skills behind them.

➤ Assessment & Next Steps

Students should complete the practice activities included in each *Math Sense 3: Focus on Analysis* lesson. Evaluate which learning goals were not met and remediate by using other resources, such as those identified in the Bridging Knowledge section. Upon successful completion, continue to the next unit.

Unit 4

FUNCTIONS

Skills-Based Questions:

1. What is the difference between a pattern and a sequence? How do you identify each one? *(Part 1)*
2. How do you write function rules for a sequence? How do you find later terms in the sequence? *(Part 2)*
3. How do you find geometric sequences? How are these sequences different than arithmetic sequences? *(Part 3)*
4. What are the domain and range? How do you evaluate functions? *(Part 4)*
5. How do you evaluate a function using a graph? How can you tell if a graph represents a function? *(Part 5)*
6. How do you determine the domain and range of a function? *(Part 6)*

Math Sense 3: Focus on Analysis: Part 1, p. 94; Part 2, p. 96; Part 3, p. 98; Part 4, p. 100; Part 5, p. 102; Part 6, p. 104

	Learning Goals:	**GED**
Knowledge Goals:	1. Describe the difference between a pattern and a sequence; explain how to identify a pattern or a sequence. *(Part 1)*	A.7.b A.7.c
	2. Describe how to write a function for a sequence and how to find the values of terms later in the sequence. *(Part 2)*	
	3. Describe the differences between geometric and arithmetic sequences; explain how to find later terms in a geometric sequence. *(Part 3)*	
	4. Explain the meaning of *domain* and *range*; describe how to evaluate a function. *(Part 4)*	
	5. Explain how to evaluate a function using a graph; describe how to tell whether or not a graph represents a function. *(Part 5)*	
	6. Explain how to determine the domain and range of a function. *(Part 6)*	
Problem-Solving Goals:	1. Find and describe patterns; figure out rules behind sequences. *(Part 1)*	A.7.b
	2. Write functions for sequences and determine nth terms using these functions. *(Part 2)*	A.7.c
	3. Determine nth terms in geometric functions. *(Part 3)*	
	4. Evaluate functions with a given value or expression. *(Part 4)*	
	5. Evaluate functions using graphs; determine whether or not a graph shows a function. *(Part 5)*	
	6. Find the domain and range of functions. *(Part 6)*	
Vocabulary Goals:	1. Define key mathematical terms.	
	2. Determine meaning of unknown vocabulary using context clues, word forms, and parts of speech.	
	3. Apply new vocabulary to mathematical tasks and discussions.	
Math Application Goals:	1. Apply knowledge of functions to solve real-life math problems. *(Parts 1–6)*	A.7.b
	2. Defend math applications and reasoning to others. *(Parts 1–6)*	A.7.c

Sample Instructional Support Strategies

➤ Bridging Knowledge

Strategy 1: ***Develop and connect background knowledge, skills, and conceptual understanding to new knowledge.***

Strategy 2: ***Use guiding questions to make connections beyond the lesson to broader math applications.***

Strategy 3: ***Use problem-solving strategies to develop, monitor, and synthesize conceptual understanding and fluency.*** *(See also Bridging Problem Solving)*

Strategy 4: ***Extend problem-solving skills and mathematical reasoning to broader math applications in life and work.*** *(See Bridging Math Application)*

1. Evaluate students' knowledge of the following mathematical skills. Utilize the chart below to develop student content knowledge as necessary.

Unit 4 **Functions**	**Part 1** Patterns and Sequences	**Part 2** Finding the *n*th Term in a Sequence	**Part 3** Working with a Geometric Sequence	**Part 4** Function Basics	**Part 5** Reading Graphs of Functions	**Part 6** Finding Domain and Range of a Function
Core Skills in Mathematics						
Scoreboost Mathematics: Algebraic Reasoning				Evaluate Functions, p. 16		
Pre-HSE Workbook: Math 2				Identifying and Evaluating Functions, p. 34		

2. The unit preview for each unit in the *Math Sense 3: Focus on Analysis* provides a list of context examples in which to apply the target math concepts as well as questions that connect to students' prior knowledge and experience (for Unit 4, p. 93). These questions also provide a bridge to broader math applications in life and work.

3. Each unit of *Math Sense 3* also provides specialized lessons that focus on problem solving and using tools with the math skills taught in the lesson. Utilize these lessons to build student knowledge in these areas.

Unit 4	Content	Page
Problem Solver	Comparing Functions Functions in the Real World	106 108
Tools	NA	

➤ Bridging Problem Solving

Strategy 1: ***Preview the problem to determine problem-solving strategies and tools and predict general solutions.***

Strategy 2: ***Develop conceptual understanding of mathematical problems using visual representations, think-alouds, and collaboration.***

Strategy 3: ***Overcome barriers to problem solving using math models, language and structural analysis, and resources.***

Strategy 4: ***Demonstrate and defend problem solving and mathematical reasoning through reverse problem solving, mental mathematics, visual representations, and peer discussions.***

1. For each part of Unit 4, orient students to the features of each mathematical concept such as the symbols, language, and structure. Students should identify the math language and symbols to determine what is being asked of them and use this information to determine which strategies to use to complete the task. Although it is not necessary for students to fill out a graphic organizer for each math problem they attempt, completing the following graphic organizer is helpful for annotating and reviewing math concepts and choosing appropriate strategies to complete tasks. The following is an example of how a student (with guidance) might fill out this graphic organizer:

	Symbols	**Language**	**Operation or Action**	**Structure**	**Strategies/Tools**
Part 1 Patterns and Sequences		patterns, sequences, finite, infinite, term, position, value	find, solve, examine, choose	patterns you can see; the sequence is a rule that lets you know what comes next	study a list of terms and look for something that repeats to find patterns, look for changes in each term to find the rule for the sequence
Part 2 Finding the *n*th Term in a Sequence	n + =	*n*th term, position, sequence, arithmetic sequence, difference	find, apply, examine, add, adjust	an arithmetic sequence is found by adding to move up in the sequence or subtracting to find the difference between terms; a function defines the rule	find the difference between numbers in a sequence; use this number as the coefficient in the function; add to adjust the function to reflect numbers in the sequence when the first value of $n = 1$
Part 3 Geometric Sequence	n ( ) ÷ $\frac{\blacksquare}{\blacksquare}$ =	geometric sequence, ratio, divide, multiply	find, divide, multiply	geometric sequence is found by multiplying to move up in a sequence or dividing to move down	set up ratios to find the next terms
Part 4 Function Basics	$f(x) = x^2 - 5$	relation, set of ordered pairs, domain, range, function	evaluate, find, substitute, simplify	a function is the rule applied to the given variable	plug in the given function into each expression; solve or simplify
Part 5 Graphs of Functions	$f(x)$ = − +	function, graph, coordinate grid, y-coordinate, x-coordinate, vertical line test, interval	evaluate, find, substitute, graph	a function has one output for each input; if a vertical line passes through more than one point, it is not a function	use a graph to evaluate a function; find the given function value and its matching coordinate; use a vertical line test to determine if it is a function
Part 6 Domain & Range	$f(x)$ = − + ≤ ≥ =	function, domain, set of real numbers, range	find, identify	domain is all real numbers except if a function value results in division by 0 or a negative in the square root symbol	find any values that when replacing the function variable results in division by 0 or the squaring of a negative number

2. Building algebraic thinking requires further exploration with math concepts in concrete ways. Here is an activity that allows students to explore how functions work by using a machine as a metaphor for a function.

Function Machine

Materials:
9 cards with function values: $f(2)$, $f(3)$, $f(6)$, etc.
3 cards with simple functions: $f(x) = 2x + 3$, etc.

Preparation:
Divide students into groups of three. Have students sit in the following formation: Student on the left sits at a 90° angle from the table, the student in the middle faces the table, and the student on the right sits at a 90° angle from the table, facing in the opposite direction of the student on the left. Give the student on the left 3 of the 9 cards with function values. Give the middle student 1 function card. Repeat this process each time students rotate into new roles (total 3 times).

Student Roles:
Student on the Left: Input: Has 3 cards each with a different *f* value [$f(2)$, $f(3)$, $f(6)$]. Hands one card to the middle student. Waits until a solution is completed by the student in the middle and checked by the student on the right. Repeats the process with the remaining 2 cards.

Student in the Middle: Function: Has 1 function card [$f(x) = 2x + 3$]. Receives a function value from the student on the left and evaluates the function using the value from the card and passes it to the student on the right.

Student on the Right: Output: Receives the evaluated function card (from the student in the middle). Checks the solution working backwards through the problem, inserting the solution into the equation to see if the resulting x value is the same as the function value from the student on the left:

Switch roles:
Each student rotates 1 chair into a new position. In the new position, each student receives new cards they need for the activity. Once they complete the activity, they rotate one more time so that each student has a chance in each of the roles.

Extension:
Create functions for real-life examples. For example, the input can be number of gallons of gas, the function can be $f(x) = 20(x)$, with 20 equal to the mpg and x the number of gallons. The output then would show *the distance* a vehicle that gets 20 mpg can go on 10 gallons of gas.

3. Make the above activity collaborative by having the student checking the solution share his/her methods and calculations with the other two students. Encourage students to refer to their math vocabulary lists (p. 161) and to use the discourse prompts (p. 159).

4. The above activity requires that the student on the right use backwards problem solving to check the solution from the student in the middle. In this way students learn the valuable skills of evaluating their work and strategizing how to fix their work when the result is incorrect.

➢ Bridging Vocabulary

Strategy 1: ***Identify the component parts and usage of new words to interpret their meanings.***

Strategy 2: ***Use context clues to interpret new words.***

Strategy 3: ***Utilize vocabulary-building resources.***

Strategy 4: ***Build a deeper knowledge of words through math application tasks and collaborative discussions.***

Strategy 5: ***Memorize words through repetitive study such as using flashcards (digital or print) and notes.***

1. First, present the shortest form of the word, referred to in this text as the "base word" in the case of academic words and some subject-specific terms. Follow the base form with other commonly used word forms (if available). Examine prefixes and suffixes and their impact on word meaning and part of speech.
2. Read the word as used in the context of the text and discuss possible meanings given context clues and word form.
3. Have students find (electronically or in print) the definition or translation of the base form and, if different, the form used in context and note these definitions in the space provided for future reference and study.
4. Gradually build deeper knowledge of the word by having students use the word in a sentence frame, guided discussion, and an original sentence within a mathematical context.

Sentence Frame:	*In a **function**, there is an ____________________, ____________________, and ____________________.*
Guided Discussion:	*Give an example of how a **function** works.*
Original Math Sentence:	____

Encourage students to use these words in math applications and collaborative discussions such as the task described in Bridging Problem Solving Strategy 4.

5. The high volume of mathematical terminology requires repeat exposure to the words over time. Word walls, intentionally including the words in questions to students and when eliciting responses from them, and explicit reminders to use the vocabulary in verbal tasks provide built-in reinforcement. However, this is often not enough so it is important that students learn ways to study words independently. Flashcards or websites that offer repetitive vocabulary practice are excellent ways for students to do this. Students may also use their notes, however, they will need to do repetitive activities, similar to flashcard practice, and not simply read and reread their notes.

➤ Bridging Math Application

Strategy 1: ***Prepare for math applications by identifying the problem type and the problem-solving strategies and tools.***

Strategy 2: ***Organize the problem using visual, symbolic, and written representations.***

Strategy 3: ***Overcome barriers to problem solving using math models, language and structural analysis, and resources.***

Strategy 4: ***Demonstrate and defend problem-solving application and mathematical reasoning through reverse problem solving, mental mathematics, visual representations, and peer discussions.***

1. Each part of this lesson lends itself to a variety of math application tasks that allow students to synthesize, apply, or extend their mathematical knowledge and skills. Whichever math application task you choose, be sure to orient students to the problem type and the problem-solving strategies and tools they may utilize. The following is an example of student directions for a math application task that synthesizes the concepts developed in this lesson.

Body Proportions

A human being is an amazing entity. Each human being is completely unique. However, is it possible that some characteristics are the same for most human beings? A simple online search with the search term "human body proportions" will give you information that says that most human bodies have the same proportions for certain body parts. Is this true?

In a small group, gather the following data from 5 individuals to determine if certain body parts are proportional to other body parts. Measure each pair of body parts in centimeters* or inches.

1. Head length (tip of chin to top of head) *and* lower leg length (below knee to bottom of heel)
2. Head length *and* overall height
3. Overall height *and* arm length (from armpit to finger tips)
4. Foot length *and* forearm (from inside elbow to wrist)

Given your measurements of 5 individuals, which of the 4 comparisons show similar proportions across all 5 individuals? 4 out of 5?

Create a function for each proportion that you discovered was consistent for at least 4 out of the 5 people you measured.

Test your function by measuring each of the following body parts from someone you have not measured yet, and predicting the measurement of the proportional body part.

1. Head length [predict: lower leg length]
2. Head length [predict: height]
3. Height [predict: arm length]
4. Foot length [predict: forearm length]

Test your predictions by measuring the actual body parts and comparing the actual measurements to your predicted ones. Were your predictions good estimates? Do you believe that most human beings have these same proportions? Why or why not?

**Hint:* Working with centimeters will result in fewer fractions. You can always convert centimeters to inches later.

Before engaging in problem solving, have students analyze the directions to determine what is being asked of them and use this information to determine which strategies to use to complete the task. The following example provides a glimpse of how this may look.

Directions	Operation or Action *Signal Words*	Strategies and Tools
Gather the following data from 5 individuals to determine if certain body parts are proportional to other body parts. Measure each pair of body parts in centimeters or inches.	gather, measure	Create a chart to record the measurements for the 5 people
Given your measurements of 5 individuals, which of the 4 comparisons show similar proportions across all 5 individuals? 4 out of 5?		

2. Students will need to organize how to complete the above task. Their analysis of the directions is a good start. They will also need a step-by-step plan for completing the task and somewhere to record their notes and work. Here is one example of how this may look:

Name	Head	Lower Leg	Head	Height	Height	Arm	Foot	Forearm
1.								
2.								
3.								
4.								
5.								
Proportion:	**yes/no**		**yes/no**		**yes/no**		**yes/no**	
Function:								

Testing Functions

Gray : Measure   White : Predict using function

Name	Head	Lower Leg	Head	Height	Height	Arm	Foot	Forearm

3. Students can collaborate with others to organize and complete the task. They can use the lessons in Unit 4 for reference as they create functions and test them.

4. As in the Bridging Problem Solving activity earlier in this lesson, students can use backwards problem solving to check their calculations in the above activity. Once they enter an input value into the function they created and get the output value, they can turn around and plug the output value into the function and find the input value. The final question in this activity allows students to use the evidence from their research to support or disclaim the hypothesis that most human beings have the same proportions between the pairs of body parts they measured.

➢ Assessment & Next Steps

Students should complete the practice activities included in each Math Sense 3: Focus on Analysis lesson. Evaluate which learning goals were not met and remediate by using other resources, such as those identified in the Bridging Knowledge section. Upon successful completion, continue to the next unit.

Unit 5

POLYNOMIALS AND RATIONAL EXPRESSIONS: Section 1

Skills-Based Questions:

1. What are polynomials and how do you simplify polynomial expressions? *(Part 1)*
2. What are monomials and how do you multiply them together or with an expression? *(Part 2)*
3. What are binomials and how do you multiply them together? *(Part 3)*
4. How do you factor out a monomial? *(Part 4)*
5. How do you factor trinomials? *(Part 5)*

Math Sense 3: Focus on Analysis: Part 1, p. 114; Part 2, p. 116; Part 3, p. 118; Part 4, p. 120; Part 5, p. 122

	Learning Goals:	GED
Knowledge Goals:	1. Explain the meaning of polynomial and describe how to simplify polynomial expressions. *(Part 1)*	A.1.d
	2. Define *monomial* and describe how to multiply monomials together or with an expression. *(Part 2)*	A.1.f
	3. Define *binomial* and describe how to use FOIL to multiply binomials together. *(Part 3)*	
	4. Explain the process for factoring out a monomial. *(Part 4)*	
	5. Describe how to factor trinomials using the idea behind FOIL. *(Part 5)*	
Problem-Solving Goals:	1. Simplify polynomial expressions by using the distributive property and combining like terms. *(Part 1)*	A.1.d
	2. Multiply monomials together or with an expression. *(Part 2)*	A.1.f
	3. Multiply binomials using FOIL. *(Part 3)*	
	4. Factor out a monomial and rewrite the expression as a product. *(Part 4)*	
	5. Factor trinomials using information from FOIL. *(Part 5)*	
Vocabulary Goals:	1. Define key mathematical terms.	
	2. Determine the meaning of unknown vocabulary using context clues, word forms, and parts of speech.	
	3. Apply new vocabulary to mathematical tasks and discussions.	
Math Application Goals:	1. Apply knowledge of polynomials to solve real-life math problems. *(Parts 1–5)*	A.1.d
	2. Defend math applications and reasoning to others. *(Parts 1–5)*	A.1.f

Sample Instructional Support Strategies

➢ Bridging Knowledge

Strategy 1: ***Develop and connect background knowledge, skills, and conceptual understanding to new knowledge.***

Strategy 2: ***Use guiding questions to make connections beyond the lesson to broader math applications.***

Strategy 3: ***Use problem-solving strategies to develop, monitor, and synthesize conceptual understanding and fluency.*** *(See also Bridging Problem Solving)*

Strategy 4: ***Extend problem-solving skills and mathematical reasoning to broader math applications in life and work.*** *(See Bridging Math Application)*

1. Evaluate students' knowledge of the following mathematical skills. Utilize the chart below to develop student content knowledge as necessary.

Unit 5 **Polynomials and Rational Expressions, Section 1**	**Part 1** Simplify Polynomial Expressions	**Part 2** Multiply an Expression by a Monomial	**Part 3** Multiply Binomials Using FOIL	**Part 4** Factor Out a Monomial	**Part 5** Use FOIL to Factor Trinomials
Core Skills in Mathematics	Unit 4, Lesson 2: Solving Equations (p. 64)				
Scoreboost Mathematics: Algebraic Reasoning	Write and Solve Equations (p. 14)				
Pre-HSE Workbook: Math 2	Solving Multi-Step Equations (p. 20)				

2. The unit preview for each unit in the *Math Sense 3: Focus on Analysis* provides a list of context examples in which to apply the target math concepts as well as questions that connect to students' prior knowledge and experience (for Unit 5, p. 113). These questions also provide a bridge to broader math applications in life and work.

3. Specialized lessons on problem solving and using tools are not provided in this unit.

➤ Bridging Problem Solving

Strategy 1: ***Preview the problem to determine problem-solving strategies and tools and predict general solutions.***

Strategy 2: ***Develop conceptual understanding of mathematical problems using visual representations, think-alouds, and collaboration.***

Strategy 3: ***Overcome barriers to problem solving using math models, language and structural analysis, and resources.***

Strategy 4: ***Demonstrate and defend problem solving and mathematical reasoning through reverse problem solving, mental mathematics, visual representations, and peer discussions.***

1. For Parts 1–5 of Unit 5, orient students to features of each mathematical concept such as the symbols, language, and structure. Students should identify the math language and symbols to determine what is being asked of them and use this information to determine which strategies to use to complete the task. Although it is not necessary for students to fill out a graphic organizer for each math problem they attempt, completing the following graphic organizer is helpful for annotating and reviewing math concepts and choosing appropriate strategies to complete tasks. The following is an example of how a student (with guidance) might fill out this graphic organizer:

	Symbols	**Language**	**Operation or Action**	**Structure**	**Strategies/Tools**
Part 1 Simplify Polynomials	x y n ( ) ■/■ + − = x^2 $3x$	terms, constant, like terms, unlike terms, coefficient, polynomial, monomial, binomial, trinomial, distributive property	simplify, combine, add, subtract	like terms have the same variable raised by the same power but can have different coefficients; unlike terms have different terms or the same term raised by a different power; polynomial has one or more terms	use the distributive property to group like terms, combine by adding or subtracting; write a simplified expression
Part 2 Multiply by a Monomial	x y n ( ) • ■/■ + − = x^2 $3x$	monomial, coefficient, variable, rules of exponents,	multiply	rules of exponents state that when you multiply exponents keep the base the same and add the power	use the rules of exponents and the distributive property to multiply monomials together or with an expression
Part 3 Multiply Binomials Using FOIL	x y n ( ) • ■/■ + − = x^2 $3x$	binomial, FOIL, parentheses, like terms	multiply, combine, simplify	use FOIL (first, outer, inner, last) to multiply binomials	FOIL helps keep track of which terms to multiply when
Part 4 Factor Out a Monomial	x y n ( ) • ■/■ + − = x^2 $3x$	factors, factoring, greatest common factor (GCF), factor out	factor, divide	factoring an expression means to reduce terms by dividing each term by the GCF, grouping the remaining terms within parentheses and putting the GCF outside	find the GCF of the terms, factor it out and rewrite the expression; when dividing an exponent, subtract the power
Part 5 Use FOIL to Factor Trinomials	x y n ( ) • ■/■ + − = x^2 $3x$	factor, FOIL, binomial factors, trinomial, product, coefficient	factor, simplify, add, multiply, find, list	the middle term's coefficient in a trinomial is a pair of factors from the last term added together	look for a pair of factors from the last term that add up to the coefficient in the middle term

2. Building algebraic thinking requires further exploration with math concepts in concrete ways. Here is an activity that allows students to explore how to work with polynomials.

Materials:
Polynomial cards: One card for each student with either a monomial, binomial, or trinomial expression* on it; make an equal number of cards for each type of polynomial if possible. (Students will form groups of 3 later on, each with a different type of polynomial)
Posters: Each stating the name of a type of polynomial: monomial, binomial, or trinomial
Binomial cards: One additional binomial card for each team of 3. (Pass these out after demonstrating how to use FOIL to multiply binomials)

Preparation:
Hang up the posters in three different areas of the room. Explain the meaning of the word roots mono–, bi–, and tri–. Show the signs in the room and have students listen and repeat the name of each polynomial. Have students match the roots you discussed with each of the names.

Polynomial Line Up:
Pass out one polynomial card to each student. Tell them to read their card and decide if the expression is a monomial, binomial, or trinomial. (Since you did not discuss the definitions of these, students will need to use their knowledge of the root words and guess which one they have.) Once they decide which they have, have them gather by the correct poster. When they are in a group by their poster, students should compare their cards and decide on a definition for the word. These definitions are shared with the class and adjusted for clarity and accuracy.

Working with Polynomials:
Have students make teams of 3 with each member coming from a group with a different type of polynomial expression. Describe to the class how to multiply a binomial with a monomial. In their teams, have students use their binomial card and multiply it by their monomial card. Each group will go to the board to show the steps and the result. Next, show students how to use FOIL to multiply two binomials, then pass out an additional binomial expression card to each team. Have students multiply this additional card to their existing binomial. Again, teams take turns going up to the board and showing their methods and result. Finally, describe to students how to factor a trinomial using the principles behind the FOIL method. Have them factor their trinomial card and demonstrate at the board.

*For this exploration activity, make these expressions simple with numbers that work well with each other.

3. The collaborative activity allows students to help each other as they explore the math concepts and develop math skills. Once these exploration activities become a norm in the classroom, students' tolerance for ambiguity increases, allowing them to take risks in problem solving. Furthermore, the exploratory nature of this type of activity sets up a low stakes environment where students, especially ELLs, can take risks using the English language both for general conversation and for math discussions. Encourage students to have their math vocabulary references at hand.

4. Each team of learners in the above activity is required to demonstrate their problem solving on the board. This should be considered as an opportunity to further develop skills, not a demonstration of proficiency. Encourage polite discussion around the methods used and the subsequent results. Student mistakes on the board should be treated as opportunities for the whole class to learn. A gracious attitude toward their struggles creates a class culture where risks will be taken and confidence bolstered.

➢ Bridging Vocabulary

Strategy 1: Identify the component parts and usage of new words to interpret their meanings.

Strategy 2: Use context clues to interpret new words.

Strategy 3: Utilize vocabulary-building resources.

Strategy 4: Build a deeper knowledge of words through math application tasks and collaborative discussions.

Strategy 5: Memorize words through repetitive study such as using flashcards (digital or print) and notes.

1. First, present the shortest form of the word, referred to in this text as the "base word" in the case of academic words and some subject-specific terms. Follow the base form with other commonly used word forms (if available). Examine prefixes and suffixes and their impact on word meaning and part of speech.

2. Read the word as used in the context of the text and discuss possible meanings given context clues and word form.

3. Have students find (electronically or in print) the definition or translation of the base form and, if different, the form used in context and note these definitions in the space provided for future reference and study.

4. Gradually build a deeper knowledge of the word by having students use the word in a sentence frame, guided discussion, and an original sentence within a mathematical context.

Sentence Frame:	*A **trinomial** contains* __________________ *such as* __________________.
Guided Discussion:	*How do we factor **trinomials**?*
Original Math Sentence:	____

Encourage students to use these words in math applications and collaborative discussions such as the task described in Bridging Problem Solving Strategy 4.

5. The high volume of mathematical terminology requires repeat exposure to the words over time. Word walls, intentionally including the words in questions to students and when eliciting responses from them, and explicit reminders to use the vocabulary in verbal tasks provide built-in reinforcement. However, this is often not enough so it is important that students learn ways to study words independently. Flashcards or websites that offer repetitive vocabulary practice are excellent ways for students to do this. Students may also use their notes, however, they will need to do repetitive activities, similar to flashcard practice, and not simply read and reread their notes.

➢ Bridging Math Application

Strategy 1: ***Prepare for math applications by identifying the problem type and the problem-solving strategies and tools.***

Strategy 2: ***Organize the problem using visual, symbolic, and written representations.***

Strategy 3: ***Overcome barriers to problem solving using math models, language and structural analysis, and resources.***

Strategy 4: ***Demonstrate and defend problem-solving application and mathematical reasoning through reverse problem solving, mental mathematics, visual representations, and peer discussions.***

1. By this point in the lesson, students may be asking if they will really need to work with polynomials with multiple variables in real life. Although there are many ways polynomials may be used in real life, the probability of your students using them for day-to-day tasks is quite small. However, the skills needed for working with polynomials are also skills that help them work with functions, quadratic equations, and a variety of useful formulas that have applications in day-to-day life. Furthermore, these skills open doors to a number of high-paying, high-demand careers. Here is an activity that will help student explore careers that use polynomials:

Careers That Work With Polynomials

Mastering the math skills needed to work with polynomial expressions sets you up well for a variety of high-paying, high-demand careers. Here is a list of careers that work with polynomials.

1. Economists
2. Science careers: archaeologists, astronomers, meteorologists, chemists, physicists, physical and social scientists
3. Engineering careers: aerospace, chemical, civil, electrical, environmental, mechanical, and industrial
4. Statisticians
5. Health careers: physicians, pharmacists, researchers
6. Computer careers: scientists, programmers
7. Business careers

What types of problems do they work with? With a partner, choose one of the above careers to research online. For your search term, type the name of the career and "math formulas," for example: "economist math formulas." Find an image that shows a formula with polynomials. Write the formula down. Insert pretend values for all but one of the variables in the formula. Solve for the remaining variable.

Do you think you would like a career that uses math? Discuss why or why not with a partner.

Before engaging in problem solving, have students analyze the directions to determine what is being asked of them and use this information to determine which strategies to use to complete the task. The following example provides a glimpse of how this may look.

Directions	**Operation or Action** *Signal Words*	**Strategies and Tools**
With a partner, choose one of the above careers to research online.	*choose*	Discuss with a partner and choose a career that is interesting; look up careers in a translator or dictionary if not understood
For your search term, type the name of the career and "math formulas," for example: "economist math formulas."		

2. Students will need to organize how to complete the above task. Their analysis of the directions is a good start. They will also need a step-by-step plan for completing the task and somewhere to record their notes and work. Here is one example of how this may look:

> **Career:** **Search Term:** ______________ **math formulas**
>
> Formula with polynomial:
>
> Pretend values:
>
> Solve for _________ value:
>
> Solution:

3. Students can collaborate with others to organize and complete the task. They can use the lessons in Unit 5 for reference as they attempt to solve the formula they choose.
4. Provide students with the opportunity to discuss the career they choose and the type of formula they tried. Provide some questions to guide their discussions and, as always, encourage them to use precise mathematical language.

> 1. What is it about this career interests you?
> 2. What formula did you choose? Please read the formula out loud.
> 3. Why did you choose this formula?
> 4. What kind of polynomials does it use and how do you know?
> 5. When you tried to solve the formula with your pretend values, what problems (if any) did you have?
> 6. Do you think if you had more math practice working with polynomials that you might want to pursue a career like this? Why or why not?

➤ Assessment & Next Steps

Students should complete the practice activities included in each *Math Sense 3: Focus on Analysis* lesson. Evaluate which learning goals were not met and remediate by using other resources, such as those identified in the Bridging Knowledge section. Upon successful completion, continue to the next section of this unit.

POLYNOMIALS AND RATIONAL EXPRESSIONS: Section 2

Skills-Based Questions:

1. What are rational expressions and how do you add and subtract them? *(Part 6)*
2. How do you multiply rational expressions? *(Part 7)*
3. How do you divide rational expressions? *(Part 8)*
4. How do you simplify and evaluate rational expressions and find undefined values? *(Part 9)*

Math Sense 3: Focus on Analysis: Part 6, p. 126; Part 7, p. 128; Part 8, p. 130; Part 9, p. 132

	Learning Goals:	GED
Knowledge Goals:	1. Explain the meaning of rational expressions and describe how to add and subtract them. *(Part 6)*	A.1.h
	2. Describe how to multiply rational expressions. *(Part 7)*	A.1.i
	3. Describe how to divide rational expressions. *(Part 8)*	
	4. Describe how to simplify and evaluate rational expressions; find undefined values. *(Part 9)*	
Problem-Solving Goals:	1. Add and subtract rational expressions. *(Part 6)*	A.1.h
	2. Multiply rational expressions. *(Part 7)*	A.1.i
	3. Divide rational expressions. *(Part 8)*	
	4. Simplify rational and define rational expressions; find undefined values. *(Part 9)*	
Vocabulary Goals:	1. Define key mathematical terms.	
	2. Determine the meaning of unknown vocabulary using context clues, word forms, and parts of speech.	
	3. Apply new vocabulary to mathematical tasks and discussions.	
Math Application Goals:	1. Apply knowledge of rational expressions to solve real-life math problems. *(Parts 6–9)*	A.1.h
	2. Defend math applications and reasoning to others. *(Parts 6–9)*	A.1.i

Sample Instructional Support Strategies

➢ Bridging Knowledge

Strategy 1: ***Develop and connect background knowledge, skills, and conceptual understanding to new knowledge.***

Strategy 2: ***Use guiding questions to make connections beyond the lesson to broader math applications.***

Strategy 3: ***Use problem-solving strategies to develop, monitor, and synthesize conceptual understanding and fluency.*** *(See also Bridging Problem Solving)*

Strategy 4: ***Extend problem-solving skills and mathematical reasoning to broader math applications in life and work.*** *(See Bridging Math Application)*

1. Evaluate students' knowledge of the following mathematical skills. Utilize the chart below to develop student content knowledge as necessary.

Unit 5 **Polynomials and Rational Expressions, Section 2**	**Part 6** Add and Subtract Rational Expressions	**Part 7** Multiply Rational Expressions	**Part 8** Divide Rational Expressions	**Part 9** Simplify Rational Expressions
Core Skills in Mathematics	Unit 4, Lesson 1: Evaluating Expressions (p. 60); Lesson 2: Solving Equations (p. 64)			
Scoreboost Mathematics: Algebraic Reasoning	Write and Solve Equations (p. 14)			Evaluate Expressions (p. 10)
Pre-HSE Workbook: Math 2	Solving Multi-Step Equations (p. 20)			

2. The unit preview for each unit in the *Math Sense 3: Focus on Analysis* provides a list of context examples in which to apply the target math concepts as well as questions that connect to students' prior knowledge and experience (for Unit 5, p. 113). These questions also provide a bridge to broader math applications in life and work.
3. Specialized lessons on problem solving and using tools are not provided in this unit.

➢ Bridging Problem Solving

Strategy 1: ***Preview the problem to determine problem-solving strategies and tools and predict general solutions.***

Strategy 2: ***Develop conceptual understanding of mathematical problems using visual representations, think-alouds, and collaboration.***

Strategy 3: ***Overcome barriers to problem solving using math models, language and structural analysis, and resources.***

Strategy 4: ***Demonstrate and defend problem solving and mathematical reasoning through reverse problem solving, mental mathematics, visual representations, and peer discussions.***

1. For Parts 6–9 of Unit 5, orient students to the features of each mathematical concept such as the symbols, language, and structure. Students should identify the math language and symbols to determine what is being asked of them and use this information to determine which strategies to use to complete the task. Although it is not necessary for students to fill out a graphic organizer for each math problem they attempt, completing the following graphic organizer is helpful for annotating and reviewing math concepts and choosing appropriate strategies to complete tasks. The following is an example of how a student (with guidance) might fill out this graphic organizer:

	Symbols	**Words**	**Operation or Action**	**Structure**	**Strategies/Tools**
Part 6 Add and Subtract Rational Expressions	x y n x^2 $3x$ $\frac{\blacksquare}{\blacksquare}$ + − =	ratio, rationale expression, polynomials, numerator, denominator, lowest common denominator	add, subtract, simplify, find	a rational expression compares two polynomials; is written as a fraction; needs like denominators to add or subtract; find least common denominator	use fraction rules for adding and subtracting with rational expressions; fractions with like denominators add or subtract numerator and keep denominator the same
Part 7 Multiply Rational Expressions	x y n x^2 $3x$ $\frac{\blacksquare}{\blacksquare}$ + − • ( ) =	rational expressions, cancel, numerator, denominator, monomials, binomials, factoring, factors	cancel, multiply, simplify	use common factors to cancel; multiply numerators and denominators across; factor trinomials	use fraction rules for multiplying with rational expressions; cancel when possible
Part 8 Divide Rational Expressions	x y n x^2 $3x$ $\frac{\blacksquare}{\blacksquare}$ + − • ( ) =	division, fractions, divisor, cancel, expression, factor, numerator, denominator	multiply, combine, simplify	flip the numerator and denominator of the divisor and change division to multiplication	use fraction rules for dividing with rational expressions, cancel when possible
Part 9 Simplify Rational Expressions	x y n x^2 $3x$ $\frac{\blacksquare}{\blacksquare}$ + − • ( ) =	simplify, rational expressions, numerator, denominator, factor cancel, distributive property	simplify, factor out, factor, cancel, find, evaluate	evaluate rational expressions by first simplifying the expression and then substituting the value	follow the step-by-step method for evaluating a rational expression: factor, simplify, multiply, substitute, simplify

2. Building algebraic thinking requires further exploration with math concepts in concrete ways. Remember those polynomial expression cards? Here is an activity that allows students another opportunity to work with them, this time within rational expressions.

Rational Numbers

Materials:
Polynomial cards: One card for each student with either a monomial, binomial, or trinomial expression* on it. (Reuse from the polynomial activity in the previous lesson.)
String: One string per group, cut to approximately 6 inches. (This will be the fraction bar.)

Preparation:
Review the rules of working with fractions. Have students take notes and/or post the rules somewhere visible in the room.

Forming Rational Numbers:
Separate students into pairs. Pass out one string to each pair. Give one polynomial card to each student. Have each pair lay one student's expression card above the other student's on the table. Then have them put their string horizontally between the two cards. Ask students what they think they just created. If they have no idea, explain that the string is a fraction bar, then ask the question again. Once the idea of a fraction is revealed, remind students that fractions represent a ratio. Then write the word "rational" on the board. What word do they recognize within it? (Ratio!) Explain to students that what they created is a special fraction, or ratio, called a rational expression. Elicit from students what a rational expression must have in the numerator and denominator. Together, write a definition for a rational expression.

Working with Rational Numbers:
Next, explain the process of adding or subtracting two rational expressions. Refer students to the rules regarding performing operations on fractions. Have two pairs of students form a team of four students. Have them set up the rational expressions they created as pairs into an equation in which they will either add or subtract one rational expression from the other. Ask volunteer teams to come to the board and demonstrate their methods and results. Next describe how to multiply rational expressions. Again have teams put their rational expressions into an equation in which they will multiply the expressions together. Volunteers can then come to the board to demonstrate and the class can discuss the methods and results. Finally, repeat the process by having students divide one rational expression from the other. Point them toward the fraction rules so they see that they will actually be multiplying again.

*For this exploration activity, make these expressions simple with numbers that work well with each other.

3. Like the previous lesson on polynomials, this collaborative activity allows students to help each other as they explore the math concepts and develop math skills. Once these exploration activities become a norm in the classroom, students' tolerance for ambiguity increases, allowing them to take risks in problem solving. Furthermore, the exploratory nature of this type of activity sets up a low-stakes environment where students, especially ELLs, can take risks using the English language both for general conversation and for math discussions. Encourage students to have their math vocabulary references at hand.

4. Since the math in this lesson is quite challenging for an exploratory activity, instead of having all teams demonstrate their work on the board have volunteers come up. Remember that this activity should be considered an opportunity to further develop skills, not a demonstration of proficiency. As such, encourage polite discussion around the methods used and the subsequent results. Student mistakes on the board should be treated as opportunities for the whole class to learn. A gracious attitude toward their struggles creates a class culture where risks will be taken and confidence bolstered.

➢ Bridging Vocabulary

Strategy 1: ***Identify the component parts and usage of new words to interpret their meanings.***

Strategy 2: ***Use context clues to interpret new words.***

Strategy 3: ***Utilize vocabulary-building resources.***

Strategy 4: ***Build a deeper knowledge of words through math application tasks and collaborative discussions.***

Strategy 5: ***Memorize words through repetitive study such as using flashcards (digital or print) and notes.***

1. First, present the shortest form of the word, referred to in this text as the "base word" in the case of academic words and some subject-specific terms. Follow the base form with other commonly used word forms (if available). Examine prefixes and suffixes and their impact on word meaning and part of speech.
2. Read the word as used in the context of the text and discuss possible meanings given context clues and word form.
3. Have students find (electronically or in print) the definition or translation of the base form and, if different, the form used in context and note these definitions in the space provided for future reference and study.
4. Gradually build deeper knowledge of the word by having students use the word in a sentence frame, guided discussion, and an original sentence within a mathematical context.

Sentence Frame:	***Rational expressions*** *are written as a* ______________ *with* ______________ *and* ______________.
Guided Discussion:	*How do we work with* ***rational expressions****?*
Original Math Sentence:	____

Encourage students to use these words in math applications and collaborative discussions such as the task described in Bridging Problem Solving Strategy 4.

5. The high volume of mathematical terminology requires repeat exposure to the words over time. Word walls, intentionally including the words in questions to students and when eliciting responses from them, and explicit reminders to use the vocabulary in verbal tasks provide built-in reinforcement. However, this is often not enough so it is important that students learn ways to study words independently. Flashcards or websites that offer repetitive vocabulary practice are excellent ways for students to do this. Students may also use their notes, however, they will need to do repetitive activities, similar to flashcard practice, and not simply read and reread their notes.

➢ Bridging Math Application

Strategy 1: ***Prepare for math applications by identifying the problem type and the problem-solving strategies and tools.***

Strategy 2: ***Organize the problem using visual, symbolic, and written representations.***

Strategy 3: ***Overcome barriers to problem solving using math models, language and structural analysis, and resources.***

Strategy 4: ***Demonstrate and defend problem-solving application and mathematical reasoning through reverse problem solving, mental mathematics, visual representations, and peer discussions.***

1. Each part of this lesson lends itself to a variety of math application tasks that allow students to synthesize, apply, or extend their mathematical knowledge and skills. Whichever math application task you choose, be sure to orient students to the problem type and the problem-solving strategies and tools they may utilize. The following is an example of student directions for a math application task that synthesizes the concepts developed in this lesson.

Getting the Job Done

You are a busy parent attending both work and school. You can't seem to find enough time to finish all the household cleaning each week. You realize you need help and decide to have your children help you. It usually takes you 3 hours to clean the whole house but you figure that with your two kids helping you it done in much less time. After a couple weeks of training the kids, you realize that your oldest child takes about 1 hour longer than you if he works by himself, and your youngest child takes twice as long as you. How quickly will you be able to get the housecleaning done if the three of you work together?

Before engaging in problem solving, have students analyze the directions to determine what is being asked of them and use this information to determine which strategies to use to complete the task. The following example provides a glimpse of how this may look.

Directions	**Operation or Action** *Signal Words*	**Strategies and Tools**
How quickly will you be able to get the housecleaning done if the three of you work together?	Quickly = time; done = work	Use a work formula to solve the problem: $w = r \cdot t$; put in combined rate and work = 1; solve for t.

2. Students will need to organize how to complete the above task. Their analysis of the directions is a good start. They will also need a step-by-step plan for completing the task and somewhere to record their notes and work. Here is one example of how this may look:

Facts	**Calculations**	**Solution**
It usually takes you 3 hours to clean the whole house by yourself		
The oldest child takes 1 hour longer		
The youngest child takes 2 hours longer		
How quickly are the three of you able to get the housecleaning done?		

3. Students can collaborate with others to organize and complete the task. They can use the lessons in Unit 5 for reference as they create equations and check them.

4. Estimation is a great way to check if your calculations are correct. Have students think about the word problem above and use mental math to figure out how quickly each of the kids could finish the project. Since the percentages are in increments of 25, thinking in terms of quarters per dollar may be helpful. If you add 75% to 50% it is obvious that the kids can finish the cleaning more quickly than the parent can, therefore, the answer should reflect this.

➢ Assessment & Next Steps

Students should complete the practice activities included in each *Math Sense 3: Focus on Analysis* lesson. Evaluate which learning goals were not met and remediate by using other resources, such as those identified in the Bridging Knowledge section. Upon successful completion, continue to the next unit.

Unit 6

QUADRATIC EQUATIONS

Skills-Based Questions:

1. What is a quadratic equation and how do you solve simple quadratic equations? *(Part 1)*
2. How do solve quadratic equations using factoring? *(Part 2)*
3. What is a trinomial square and how do you solve quadratic equations with trinomial squares? *(Part 3)*
4. How do you interpret and solve word problems with quadratic equations? *(Part 4)*

Math Sense 3: Focus on Analysis: Part 1, p. 138; Part 2, p. 142; Part 3, p. 144; Part 4, p. 148

	Learning Goals:	GED
Knowledge Goals:	1. Explain the meaning of quadratic equation and describe how solve simple quadratic equations. *(Part 1)*	A.4.a A.4.b
	2. Describe how to solve quadratic equations using factoring. *(Part 2)*	
	3. Define a trinomial square; explain how to solve quadratic equations with trinomial squares. *(Part 3)*	
	4. Describe how to interpret and solve word problems with quadratic equations. *(Part 4)*	
Problem-Solving Goals:	1. Add and subtract rational expressions. *(Part 1)*	A.4.a
	2. Solve quadratic equations by factoring. *(Part 2)*	A.4.b
	3. Solve quadratic equations that use trinomial squares by completing the square. *(Part 3)*	
	4. Interpret and solve word problems with quadratic equations. *(Part 4)*	
Vocabulary Goals:	1. Define key mathematical terms.	
	2. Determine the meaning of unknown vocabulary using context clues, word forms, and parts of speech.	
	3. Apply new vocabulary to mathematical tasks and discussions.	
Math Application Goals:	1. Apply knowledge of quadratic equations to solve real-life math problems. *(Parts 1–4)*	A.4.a
	2. Defend math applications and reasoning to others. *(Parts 1–4)*	A.4.b

Sample Instructional Support Strategies

➤ Bridging Knowledge

Strategy 1: ***Develop and connect background knowledge, skills, and conceptual understanding to new knowledge.***

Strategy 2: ***Use guiding questions to make connections beyond the lesson to broader math applications.***

Strategy 3: ***Use problem-solving strategies to develop, monitor, and synthesize conceptual understanding and fluency.*** *(See also Bridging Problem Solving)*

Strategy 4: ***Extend problem-solving skills and mathematical reasoning to broader math applications in life and work.*** *(See Bridging Math Application)*

1. Evaluate students' knowledge of the following mathematical skills. Utilize the chart below to develop student content knowledge as necessary.

Unit 6 **Quadratic Equations**	**Part 1** Quadratic Equation Basics	**Part 2** Solve Quadratic Equations by Factoring	**Part 3** Completing the Square	**Part 4** Solving Word Problems with Quadratic Equations
Core Skills in Mathematics	Unit 4, Lesson 1: Evaluating Expressions (p. 60); Lesson 2: Solving Equations (p. 64)			
Scoreboost Mathematics: Algebraic Reasoning	Write and Solve Equations (p. 14); Use FOIL with Quadratic Equations (p. 20)			Evaluate Expressions (p. 10)
Pre-HSE Workbook: Math 2	Solving Multi-Step Equations (p. 20)			

2. The unit preview for each unit in the *Math Sense 3: Focus on Analysis* provides a list of context examples in which to apply the target math concepts as well as questions that connect to students' prior knowledge and experience (for Unit 6, p. 137). These questions also provide a bridge to broader math applications in life and work.

3. Each unit of *Math Sense 3* also provides specialized lessons that focus on problem solving and using tools with the math skills taught in the lesson. Utilize these lessons to build student knowledge in these areas.

Unit 6	**Content**	**Page**
Problem Solver	NA	
Tools	Solve Quadratic Equations Using Graphs Use the Quadratic Formula	140 146

➤ Bridging Problem Solving

Strategy 1: ***Preview the problem to determine problem-solving strategies and tools and predict general solutions.***

Strategy 2: ***Develop conceptual understanding of mathematical problems using visual representations, think-alouds, and collaboration.***

Strategy 3: ***Overcome barriers to problem solving using math models, language and structural analysis, and resources.***

Strategy 4: ***Demonstrate and defend problem solving and mathematical reasoning through reverse problem solving, mental mathematics, visual representations, and peer discussions.***

1. For each part of Unit 6, orient students to the features of each mathematical concept such as the symbols, language, and structure. Students should identify the math language and symbols to determine what is being asked of them and use this information to determine which strategies to use to complete the task. Although it is not necessary for students to fill out a graphic organizer for each math problem they attempt, completing the following graphic organizer is helpful for annotating and reviewing math concepts and choosing appropriate strategies to complete tasks. The following is an example of how a student (with guidance) might fill out this graphic organizer:

	Symbols	**Language**	**Operation or Action**	**Structure**	**Strategies/Tools**
Part 1 Quadratic Equation Basics	x^2 y^2 $\frac{\square}{\square}$ + − = {} $\sqrt{\ }$	quadratic equation, squared, raised to a power of 2, exponent, power, base, square root, braces	multiply, find	a quadratic equation has a variable that is squared	find square roots to solve a quadratic equation
Part 2 Solve Quadratic Equations by Factoring	x^2 + − = {}	division, fractions, divisor, cancel, expression, factor, numerator, denominator	multiply, combine, simplify	factoring means to rewrite a sum as a product; sum goes in parentheses and multiplier goes outside; have equation equal 0.	solve for x by factoring; set factors to 0 and solve for x.
Part 3 Completing the Square	x^2 + − = {} $\sqrt{\ }$	trinomial square, perfect square, factor, binomial factors, completing the square	solve, combine, divide, square, add, subtract, factor, isolate	to find the perfect square make an open place in the trinomial (combine terms without a variable), divide middle term by 2 and square the result, fill in the space with the result	solve the equation using the perfect square as the third term of a trinomial
Part 4 Solving Word Problems with Quadratic Equations	x^2 + − = {} $\sqrt{\ }$	word problems, quadratic expression, quadratic equation, formula, extraneous solution	substitute, solve, add, subtract, multiply, divide	substitute the given values into a given formula or write an equation with the values	use FOIL to write quadratic equations; watch for extraneous solutions; use common sense to choose the correct solution

2. Building algebraic thinking requires further exploration with math concepts in concrete ways. Here is an activity that allows students to explore how to work with polynomials.

Solving Quadratic Equations

First, review with students how to solve basic equations. Have students write down rules in their notebooks or create a poster that you can hang up in the classroom for reference. Next, provide a word problem that may be solved by a quadratic equation using each of the following methods: graphing, factoring, the quadratic equation, and completing the square. Some examples of word problems can be found on page 148 of *Math Sense 3: Focus on Analysis*. Demonstrate, with as much input from students as possible, how to solve the problem using each of these methods.

Project a new word problem that may be solved by a quadratic equation using each of the methods. Choose a problem that provides a formula. Set a timer for 7 minutes and have pairs of students work together to solve the word problem using graphing. Reset the timer for 7 minutes and have pairs of students solve the same word problem using factoring. Repeat this process until students have used each method to solve the same word problem. At the end of the activity, have students vote on which method they preferred and share why they preferred it.

3. Providing math models is a great way to support students as they develop their problem-solving skills. The above activity uses the same word problem and solves it in four different ways. After completing this activity, a reference copy of this problem worked through using each method would be a great model for students as they work through other quadratic equations.

4. At the end of the above activity, students are asked discuss what method they prefer for problem solving. The question asks them to explain why they prefer that method and gives them an opportunity to think through the process and match it to their current skills. However, let students know that learning all of the strategies gives them more tools to work with on the GED test and that they will see that certain problems lend themselves to using certain methods.

➢ Bridging Vocabulary

Strategy 1: Identify the component parts and usage of new words to interpret their meanings.

Strategy 2: Use context clues to interpret new words.

Strategy 3: Utilize vocabulary-building resources.

Strategy 4: Build a deeper knowledge of words through math application tasks and collaborative discussions.

Strategy 5: Memorize words through repetitive study such as using flashcards (digital or print) and notes.

1. First, present the shortest form of the word, referred to in this text as the "base word" in the case of academic words and some subject-specific terms. Follow the base form with other commonly used word forms (if available). Examine prefixes and suffixes and their impact on word meaning and part of speech.
2. Read the word as used in the context of the text and discuss possible meanings given context clues and word form.
3. Have students find (electronically or in print) the definition or translation of the base form and, if different, the form used in context and note these definitions in the space provided for future reference and study.
4. Gradually build deeper knowledge of the word by having students use the word in a sentence frame, guided discussion, and an original sentence within a mathematical context.

Sentence Frame:	*A **power** contains a __________________, which is __________________ and a __________________, which is __________________.*
Guided Discussion:	*What does a **power** mean? When would we use powers?*
Original Math Sentence:	____________________________________

Encourage students to use these words in math applications and collaborative discussions such as the task described in Bridging Problem Solving Strategy 4.

5. The high volume of mathematical terminology requires repeat exposure to the words over time. Word walls, intentionally including the words in questions to students and when eliciting responses from them, and explicit reminders to use the vocabulary in verbal tasks provide built-in reinforcement. However, this is often not enough so it is important that students learn ways to study words independently. Flashcards or websites that offer repetitive vocabulary practice are excellent ways for students to do this. Students may also use their notes, however, they will need to do repetitive activities, similar to flashcard practice, and not simply read and reread their notes.

➢ Bridging Math Application

Strategy 1: Prepare for math applications by identifying the problem type and the problem-solving strategies and tools.

Strategy 2: Organize the problem using visual, symbolic, and written representations.

Strategy 3: Overcome barriers to problem solving using math models, language and structural analysis, and resources.

Strategy 4: Demonstrate and defend problem-solving application and mathematical reasoning through reverse problem solving, mental mathematics, visual representations, and peer discussions.

1. Each part of this lesson lends itself to a variety of math application tasks that allow students to synthesize, apply, or extend their mathematical knowledge and skills. Whichever math application task you choose, be sure to orient students to the problem type and the problem-solving strategies and tools they may utilize. The following is an example of student directions for a math application task that synthesizes the concepts developed in this lesson.

Community Garden Planning

Your class has an opportunity to plant a community garden outside your school. You are given a rectangular plot with a length of 20 feet and a width of 16 feet. You want to make a walkway around the garden on the inside border of the plot. The walkway will be the same width all the way around the garden (so the ratio of the garden will be the same as the dimensions of the plot). You will need to reduce the length and width of the plot to make room for this walkway. The school has already purchased dirt and plants for a garden with an area 140 square feet. If you plan to use all the dirt and plants they purchased, what will the dimensions of the actual garden be? What will the width of the walkway be? If the school purchased enough dirt to evenly cover the area of the garden with 6 inches of dirt, how many cubic yards of dirt did they purchase?

Before engaging in problem solving, have students analyze the directions to determine what is being asked of them and use this information to determine which strategies to use to complete the task. The following example provides a glimpse of how this may look.

Directions	**Operation or Action** *Signal Words*	**Strategies and Tools**
What will the dimensions of the actual garden be?		Set up a quadratic equation with the length and width reduced by the same amount, or x, and multiplied together to get the area of 140 square feet.
What will the width of the walkway be?		

2. Students will need to organize how to complete the above task. Their analysis of the directions is a good start. They will also need a step-by-step plan for completing the task and somewhere to record their notes and work. Here is one example of how this may look:

Problem	**Equation**	**Solution**
What will the dimensions of the actual garden be?		
What will the width of the walkway be?		
If the school purchased enough dirt to evenly cover the area of the garden with 6 inches of dirt, how many cubic yards of dirt did they purchase?		

3. Students can collaborate with others to organize and complete the task. They can use the model of solving quadratic equations as a reference. They can also reference the lessons in Unit 6 as needed.

4. Students can use backwards problem solving to check their calculations in the above activity. Once they arrive at their solution, they can substitute x in the original equation for the value to see if the equation works. If the solution doesn't work, it may be that the equation itself is flawed. Have students meet with other students to discuss the algebraic reasoning behind their equations and methods they used to find their solutions. As always, encourage use of precise mathematical language.

➤ Assessment & Next Steps

Students should complete the practice activities included in each Math Sense 3: Focus on Analysis lesson. Evaluate which learning goals were not met and remediate by using other resources, such as those identified in the Bridging Knowledge section. Upon successful completion, continue to the Simulated GED Test.

APPENDIX A

Language for Specific Purposes

SIGNAL WORDS

Purpose	Language
Addition	added to, additional, all together, combined, gain of, how many in all, how many all together, how much all together, how many total, in all increase of, increased by, plus, sum total
Subtractions	change, decrease of, decreased by, difference, dropped, have left, how many left (more), how many remain (fewer), how much less (more), less than, loss of, minus, remaining, take away, taken away
Multiplication	as much as, by, each … all, increase, increased by, multiplied by, of, product of, same amount, times, triple, twice
Division	cut into, divided by, each, evenly, every, goes into, half, out of, per, into, quotient of, share equally, split into, third

DISCOURSE PROMPTS

Purpose	Language
Contribute to the Group	I think ____ because ____. I believe we should ____ because ____. My idea is that ____. Let's ____ so that ____.
Encourage Participation	What do you think? I'd like to hear what you have to say about ____. Do you have anything to add? We haven't heard from ____.
Present Evidence	An example of this is ____. ____ is an example of ____. The graph shows ____.
Disagree Politely	I see what you're saying, but ____. Another way to think about this is ____. Have you thought about ____? I think there's another way to look at that.
Paraphrase Contributions	So, do we agree that ____? I think you're saying ____. It sounds like we agree that ____. If I understand correctly, ____.
Probe Others' Contributions	Could you explain that further? What did you mean by ____? Can you give me an example of what you mean? Can you support what you're saying with evidence?

LANGUAGE FOR TEST-TAKING

Verbs	Nouns/Adjectives + Nouns			Adjectives
apply	absolute value	inequalities	quadratic equations	positive
calculate	algebra	input	quantity	negative
compare	area perimeter	length	radius	undefined
compute	arithmetic	linear expressions	range	actual
convert	bar graphs	magnitude	rational expressions	scale
create	circle graphs	mean	rational number	unknown
determine	circles	median	ratios	equivalent
display	circumference	mode	rectangles	proportional
evaluate	combinations	multiples	right prism	geometric
find	composite figures	number line	right	algebraic
graph	cone	numerical expressions	triangle	
identify	coordinate plane	output	scale factors	
identify	counting techniques	parallel lines	scatter plots	
interpret	cube roots	percents	slope	
locate	cubes	permutations	solutions	
place	cylinder	perpendicular lines	sphere	
represent	data	points	square roots	
represent	decimals	polygons	squares	
simplify	diameter	polynomials	surface	
sketch	distance	probability	system of equations	
solve	dot plots	problems	tables	
understand	factors	properties	triangles	
use	features	proportions	unit rate	
write	functions	pyramid	variable	
	geometry	Pythagorean Theorem	volume	
	histograms			

APPENDIX B

Vocabulary Development Lists

Mathematics: Math Sense 1: Focus on Operations

➢ Unit 1: Whole Numbers

Part 1: Addition	Part 2: Subtraction	Part 3: Multiplication	Part 4: Division
estimate how much/many place value regrouping total	difference between how far apart how much more	in 4 months in a year times tables	compatible numbers in 1 month length of each per week

➢ Unit 2: Decimals, Section 1

Part 1: Understanding Decimals	Part 2: Writing Decimals	Part 3: Rounding Decimals	Part 4: Comparing Decimals
place value whole numbers ten*s*/ten***ths*** decimals place values	decimal point leading zero placeholder zero	rounding whole number place value	greater than less than equal to

➢ Unit 2: Decimals Section 2

Part 5: Adding and Subtracting Decimals	Part 6: Multiplying Decimals	Part 7: Dividing Decimals
decimal point difference between estimate how far apart how many how much how much more place values regrouping total	decimal point estimate in a month in a year place values powers of ten	compatible numbers decimal point dividend divisor in a month length of each per week place values powers of ten

➢ Unit 3: Fractions, Section 1

Part 1: Relating Decimals and Fractions	Part 2: Different Forms of Fractions	Part 3: Equivalent Fractions	Part 4: Comparing Fractions	Part 5: Adding and Subtracting Like Fractions
denominator fraction numerator place value ten***s***/ten***ths***	denominator improper mixed number numerator proper remainder	cross multiply cross product denominator higher/lower terms numerator raise simplify	common denominator denominator equal to greater than less than numerator	add common denominator denominator how much faster like/unlike numerator reduce

➢ Unit 3: Fractions, Section 2

Part 6: Finding Common Denominators	Part 7: Adding and Subtracting Unlike Fractions	Part 8: Working with Whole and Mixed Numbers	Part 9: Multiplying Fractions	Part 10: Dividing Fractions	Part 11: Multiplying and Dividing with Mixed Numbers
common denominator least common multiple	common denominator unlike fractions equivalent fractions numerator simplify	common denominator equivalent fractions mixed number whole number	cancel common factors simplify	invert cancel common factors simplify reciprocal	invert mixed numbers improper fractions cancel reciprocal simplify

➢ Unit 4: Ratio, Proportion, and Percent, Section 1

Part 1: Relating Fractions and Ratios	Part 2: Writing Ratios	Part 3: Ratios in Word Problems	Part 4: Writing Proportions Part 5: Solving Problems with Proportions
compare fraction lowest terms part ratio simplify whole	ratio ratio of total units (width, length, height, minutes, etc.)	compare how many added missing information (add or subtract) quantity ratio	cross products equations equivalent fractions proportion ratio variable

➢ Unit 4: Ratio, Proportion, and Percent, Section 2

Part 6: Understanding Percents	Part 7: Decimals, Fractions, and Percents	Part 8: The Percent Equation	Part 9: Solving Percent Equations	Part 10: Two-Step Percent Problems	Part 11: Percent of Increase/ Decrease
part percent whole	decimal fraction numerator denominator percent point	equation part percent statement unknown variable whole	part percent equation variable unknown whole	percent equation variable two-step	change decrease increase original part whole

➢ Unit 5: Measurement

Part 1: Customary Units of Length	Part 2: Working with Length	Part 3: Measuring Capacity	Part 4: Measuring Weight	Part 5: Using Metric Units	Part 6: Working with Time and Temperature
customary system length feet inch mile measurement unit yard	add divide length multiply subtract unit	capacity cup fluid ounce gallon granular liquid measurement pint quart unit	ounce pound ton unit weight	centimeter decimal point place kilometer measurement meter Metric system millimeter tens unit	day hour minute month second time week year

Mathematics: Math Sense 2: Focus on Problem Solving

➢ Unit 1: Numbers and Properties

Part 1: Number Line	Part 2: Comparing and Ordering Integers	Part 3: Adding and Subtracting Signed Numbers	Part 4: Multiplying and Dividing Signed Numbers	Part 5: Powers and Roots	Part 6: Order of Operations
decimals fractions negative positive signed numbers whole numbers	equal to greater than integer less than negative positive	add negative number line signed numbers positive subtract	divide multiply negative positive signed numbers	base cube exponent perfect square power radical sign root square	brackets exponents expression fraction bar grouping order of operations parentheses powers radical sign roots value

➢ Unit 2: The Basics of Algebra, Section 1

Part 1: Expressions and Variables	Part 2: Evaluating Expressions	Part 3: Simplifying Expressions	Part 4: Negative Exponents	Part 5: Simplifying Radicals
decreased by difference divided by equation expression greater than increased by inequality less than product quotient sum times variable	expression grouping symbol negation symbol negative order of operations value variable	coefficients equivalent expressions like terms simplified variables	canceling denominator fraction bar negative exponents numerator	factor perfect square product property quotient property radicals simplify square root

➢ Unit 2: The Basics of Algebra, Section 2

Part 6: Solving Addition and Subtraction Equations	Part 7: Solving Multiplication and Division Equations	Part 8: Multistep Equations	Part 9: Equations with Separated Terms	Part 10: Equations with Parentheses	Part 11: Solving Inequalities
addition	denominator	addition	addition	distributive property	equal to
equation	division	division	division	fraction bar	greater than
inverse operations	equation	inverse operations	inverse operations	grouping symbols	greater than or equal to
numerical terms	inverse operations	multiplication	like terms	parentheses	inequalities
simplify	multiplication	subtraction	multiplication		inequality symbol
subtraction	numerator	variable	subtraction		less than
variable	reciprocal		term		less than or equal to
	variable		variable		

➢ Unit 3: Solving Problems with Algebra

Part 1: Translating Words to Equations	Part 2: Number Puzzles and Age Problems	Part 3: Solving Motion Problems	Part 4: Solving Value Problems	Part 5: Solving Work Problems	Part 6: Rewriting Formulas
addition	difference	distance	amount	denominator	equivalent formulas
chart	product	formula	how many	fraction	formula
division	quotient	how far	how much	how long	inverse operations
equation	sum	how fast	number	numerator	isolate
fact		how long	total	rate	variables
multiplication		rate	value	reciprocal	
subtraction		time	worth	time	
unknown		what speed		work	
variable					

➢ Unit 4: Geometry Basics, Section 1

Part 1: Points, Lines, and Angles	Part 2: Working with Angles	Part 3: Quadrilaterals	Part 4: Triangles	Part 5: The Pythagorean Theorem	Part 6: Similar Geometric Figures
acute horizontal intersect line line segment obtuse parallel perpendicular point ray right rotation straight vertex (vertices) vertical	complementary right angle straight angle sum supplementary	adjacent diagonal opposite parallelogram polygon, quadrilateral rectangle rhombus square trapezoid	equilateral isosceles right scalene triangle	hypotenuse legs right angle	length proportion ratio similar figures width

➢ Unit 4: Geometry Basics, Section 2

Part 7: Perimeter	Part 8: Area of Squares, Rectangles, and Parallelograms	Part 9: Area of Triangles and Trapezoids	Part 10: Circumference and Area of Circles	Part 11: Volume of Prisms and Cylinders	Part 12: Volume of Pyramids, Cones, and Spheres	Part 13: Surface Area
add distance length perimeter sides	area base height multiply parallelogram rectangle square squared surface two-dimensional figures	area base height parallelogram square trapezoid triangle	circle circumference diameter formula pi radius	cubed cylinder depth edge face height length prism three-dimensional vertex volume width	cone cubed length pyramid sphere three-dimensional	faces squared surface area three-dimensional

➢ Unit 5: Connecting Algebra and Geometry

Part 1: The Coordinate Plane	Part 2: Use Intercepts to Graph a Line	Part 3: Slope: Rise over Run	Part 4: Parallel and Perpendicular Lines	Part 5: Write the Equation of a Line	Part 6: Use Point-Slope Form	Part 7: Distance between Points
ordered pairs	linear equation	coordinate plane	fall	equation	point-slope form	absolute value
origin	*x*-intercept	rise	negative reciprocal	slope	slope	horizontal
rectangular coordinate plane	*y*-intercept	run	negative slope	slope-intercept	y-intercept	radical
x-axis		slope	parallel	*y*-axis		square
x-coordinate			perpendicular	*y*-intercept		vertical
y-axis			positive slope			*x*-axis
y-coordinate			rise			*y*-axis
			slope			

Mathematics: Math Sense 3: Focus on Analysis

➢ Unit 1: Data Analysis, Section 1

Part 1: Working with Tables	Part 2: Bar Graphs and Histograms	Part 3: Circle Graphs	Part 4: Scatter Plots	Part 5: Line Graphs	Part 6: Finding the Mean	Part 7: Median and Mode	Part 8: Weighted Averages
categories	bar graph	circle graph	correlation	approximate	average	count	mode
column	categories	combined	horizontal axis	axis	central tendency	least to greatest	outliers
compared to	data bars	decimal	negative	changes over time	mean	median	
conclusions	decrease	fraction	poll	compare	median		
data	histogram	how many	positive	data points	mode		
decrease	horizontal axis	how much	relationship	decrease	range		
difference	increase	parts	results	greater than	typical value		
headings	interval	percent	scatter plot	horizontal axis			
increase	title	sections	sets of data	increase			
row	vertical axis	total	survey	less than			
table		whole	vertical axis	line graph			
title				plot			
				vertical			

➤ Unit 1: Data Analysis, Section 2

Part 1: Reading a Box Plot	Part 2: Distribution of Data	Part 3: Making Predictions and Identifying Trends	Part 4: Understanding Correlation	Part 5: Sampling a Population	Part 6: Using Two Data Sources
box plot box-and-whisker-plot difference lower quartile maximum median minimum upper quartile	curve curve distribution histogram mean median mode normal distribution outliers range skewed (left, right) symmetric variation (low, high)	decrease increase pattern predictions remain steady trends	positive correlation negative correlation linear correlation line of best fit nonlinear correlation horizontal axis vertical axis	bias (biased) draw conclusions entire population population representative results sample skew unbiased	data source difference greater than how many how much larger less less than more smaller

➤ Unit 2: Counting and Probability

Part 1: Permutations	Part 2: Combinations	Part 3: Simple Probability	Part 4: Compound Probability	Part 5: Experimental Probability
permutations order total how many possible possibilities different ways	combination different how many possibilities possible random	chance favorable outcomes likelihood odds percent possible outcomes probability ratio	compound probability dependent independent multiply	experimental probability favorable outcomes trials prediction

➢ Unit 3: Systems of Equations and Inequalities

Part 1: Systems of Equations	Part 2: Solve Systems Using Substitution	Part 3: Solve Systems Using Elimination	Part 4: Graph Linear Equalities	Part 5: Solve Systems of Linear Equalities
brace graph system of equations variables x-axis x-coordinate y-axis y-coordinate	isolate substitution method system of equations variables	combine elimination like terms	inequality range of solutions shaded portion solution variables	system of inequalities range of solutions shading overlap

➢ Unit 4: Functions

Part 1: Patterns and Sequences	Part 2: Finding the *n*th Term in a Sequence	Part 3: Working with a Geometric Sequence	Part 4: Function Basics	Part 5: Reading Graphs of Functions	Part 6: Finding Domain and Range of a Function
finite infinite patterns position sequences term value	*n*th term position sequence arithmetic sequence difference	divide geometric sequence multiply ratio	domain function range relation set of ordered pairs	coordinate grid function graph interval vertical line test x-coordinate y-coordinate	domain function range set of real numbers

➢ Unit 5: Polynomials and Rational Expressions, Section 1

Part 1: Simplify Polynomial Expressions	Part 2: Multiply and Expression by a Monomial	Part 3: Multiply Binomials Using FOIL	Part 4: Factor Out a Monomial	Part 5: Use FOIL to Factor Trinomials
binomial coefficient constant distributive property like terms monomial polynomial terms trinomial unlike terms	coefficient monomial rules of exponents variable	binomial FOIL like terms parentheses	factor out factoring factors greatest common factor (GCF)	binomial factors coefficient factor FOIL product trinomial

➢ Unit 5: Polynomials and Rational Expressions, Section 2

Part 6: Add and Subtract Rational Expressions	Part 7: Multiply Rational Expressions	Part 8: Divide Rational Expressions	Part 9: Simplify Rational Expressions
denominator	binomials	cancel	denominator
lowest common denominator	cancel	denominator	distributive property
numerator	denominator	division	factor cancel
polynomials	factoring, factors	divisor	numerator
ratio	monomials	expression	rational expressions
rational expressions	numerator	factor	simplify
	rational expressions	fractions	
		numerator	

➢ Unit 6: Quadratic Equations

Part 1: Quadratic Equation Basics	Part 2: Solve Quadratic Equations by Factoring	Part 3: Completing the Square	Part 4: Solving Word Problems with Quadratic Equations
base	cancel	binomial factors	extraneous solution
braces	denominator	completing the square	formula
exponent, power	division	factor	quadratic equation
quadratic equation	divisor	perfect square	quadratic expression
raised to a power of 2	expression	trinomial square	word problems
square root	factor		
squared	fractions		
	numerator		

APPENDIX C

Vocabulary Development Template

1. ***Word*** or ***phrase*** (usage): symbol (if applicable)	
Definition:	
In Context:	
Sentence Frame:	
Guided Discussion:	
Original Math Sentence:	

2. ***Word*** or ***phrase*** (usage): symbol (if applicable)	
Definition:	
In Context:	
Sentence Frame:	
Guided Discussion:	
Original Math Sentence:	

3. ***Word*** or ***phrase*** (usage): symbol (if applicable)	
Definition:	
In Context:	
Sentence Frame:	
Guided Discussion:	
Original Math Sentence:	

4. ***Word*** or ***phrase*** (usage): symbol (if applicable)	
Definition:	
In Context:	
Sentence Frame:	
Guided Discussion:	
Original Math Sentence:	

5. ***Word*** or ***phrase*** (usage): symbol (if applicable)	
Definition:	
In Context:	
Sentence Frame:	
Guided Discussion:	
Original Math Sentence:	